THE LEAF BOOK

Happy Thanksgiving,
Tanner
1990

The Leaf Book

A Field Guide
to Plants of Northern California

By IDA GEARY

Illustrated with plant prints by the author

Foreword by David Cavagnaro

Published by
A. Philpott, The Tamal Land Press
Fairfax, California / 1972

Also by Ida Geary
MARIN TRAILS
published 1969; second printing 1970

Work on *The Leaf Book*
was sponsored by Marin Audubon Society
and its publication aided in part by Audubon Canyon Ranch

International Standard Book Number 0-912908-01-7
Library of Congress Catalog Card Number 78-188679

A. Philpott, The Tamal Land Press, Fairfax, Calif. 94930

Printed in the United States of America
Design and production by Arlen Philpott

If I could only put my feet on a little piece of grass . . .

IDA SARAH HELNER
Kiev, 1870
Tr. from the Russian

Contents

Foreword to The Leaf Book

People come to an appreciation of plants in all sorts of ways. Some start early in life, others late, though no matter where we have grown up plants have been an essential part of our lives since the very beginning. We have grown them, pruned them, picked their flowers, eaten their fruits, praised their beauty, or cursed them as weeds. No matter what we have thought of plants along the way, we have all been soothed uncountable times by a touch of green along a city street; we have all experienced the great relief which only wild expanses of green things growing can offer. And, what is more important, from the moment of our first breath we have breathed the oxygen put into the air untold millions of years ago, and replenished ever since, by plants. At long last we are realizing this, and many a bumper sticker now proclaims, "Have you thanked a green plant lately?" The rear end of an automobile, shrouded in smoke and noxious fumes, seems an incongruous place for such a poignant thought to find expression, but at least we have made a beginning.

It is not usually through some deep understanding of their importance, however, that we first develop our love of plants. It happens most often quite by chance, and in peculiar ways. My own fascination with plants began in the family vegetable garden when I was quite young and was renewed in earnest later when my father and I developed as a hobby the cultivation of Japanese bonsai. We used both wild and introduced trees and shrubs, and it wasn't long before we had accumulated a fair knowledge of botany.

My deepest awareness of native California plants came, however, as a result of a totally different hobby – insect collecting. Discovering bees and butterflies on certain flowers, rearing caterpillars or gall insects from specific kinds of plants, taught me almost as much about plants as about insects.

I have since met many people who have discovered the plant world by way of an interest in animals. Ida Geary is one of those people. Long before Ida ever heard of Dutchman's Pipe or Lizard-tail she was busy watching birds, learning to identify them, observing their habits. But she soon realized that birds don't exist alone. They have intimate associations with their environment, and quite often these associations involve certain plants which they frequent, upon which they feed, or in which they nest. It wasn't long before Ida was becoming a first-rate expert on the local flora. *The Leaf Book* represents a portion of the knowledge she has gained as a result of this fortunate plant affliction to which so many of us eventually fall victim.

The magnificent, delicate plant prints with which Ida Geary has illustrated this book represent also the outgrowth of a hobby which rapidly became a tool for learning the plants she encountered in the Bay Area's many wild places. In pursuing plant printing as an art form, Ida discovered that while she worked carefully with each leaf and flower, pressing the rice paper firmly around every contour, she developed what she aptly termed a "muscle memory" for each species, and her familiarity with native plants grew faster as a result.

I have felt since the beginning, and others have too who have seen Ida's work, that plant prints would be a superb tool for plant identification. Like any other medium, they are a representation, an interpretation, but nevertheless they are made from the real thing, and the forms and textures which come through in the prints are true to nature. It is as though the very essence of each species finds expression in Ida's illustrations. In addition, each print is a graceful work of art, and this alone adds another dimension to what already serves as an excellent field guide.

Learning to recognize plants comes differently to each person. Some can read this drawing better than that one, some prefer photographs, some like plants arranged by the color of their flowers while others would rather learn at the same time something about their taxonomic relationships. No matter what your preference, I believe you will find *The Leaf Book* to be a splendid new tool for identification. I believe also that you will develop, as Ida Geary herself has, a deeper love for the vast green kingdom of organisms with which we share this planet, and upon which our very survival depends. At the very least, the next time you wish to thank a green plant, you may be able to address it by name!

David Cavagnaro

Acknowledgments

"*Oh, I get by with a little help from my friends,*" the Beatles sing . . .

First, I have been helped by John Thomas Howell, curator of botany emeritus at the California Academy of Sciences. Mr. Howell, in the form of his book, *Marin Flora*, and in person has been my foremost guide and mentor.

In 1968 Mr. Howell saw my first and rudimentary collection of prints but he encouraged me to continue and he has been unfailingly generous since then – with encouragement, with letters, and with identifications of puzzling plants. I have realized since knowing Mr. Howell that a friend is someone you can take an ink-smeared plant specimen to and ask for an identification. And he will know and tell you what it is.

Again in spring of 1970 Mr. Howell went over the complete manuscript of *The Leaf Book*, and checked each page for fidelity of reproduction and accuracy of name and identifying line. I have incorporated all his changes and suggestions and I am most deeply indebted to him.

David and Maggie Cavagnaro of Stinson Beach also have seen *The Leaf Book* in process and they too have helped with identifications and often with collecting. Late in 1969 they went over the complete manuscript, refining, commenting, and suggesting deletions and inclusions. It was David's good suggestion that an identifying line be added to each picture, while Maggie has provided the drawing of Poison Oak, a plant too hazardous to print. When David was asked for a foreword he promptly produced one.

Alfonzo Molina, instructor in the science department at College of Marin, in whose plant ecology class I once seemed permanently enrolled, is another generous teacher. He too helped to collect and identify and he has seen the manuscript in one of its middle stages. Many members of his plant class – Wilma Follette, Lilian McHoul, Fay Jensen, Jane Anderson, Elizabeth Lennon, Pat Welsh, Robert Jordan, Dr. William Bortfield, Dr. Robert C. West, are a few – were good companions on field trips and interested in this project.

Field trips were a necessary part of this endeavor and I have enjoyed occasions with the California Native Plant Society and the repeated fellowship of members of Marin Audubon Society. Some of them are Malcolm and Laura Smith, Gwynne Peticolas, Elizabeth Terwilliger, Helen Pratt, Paula Dawson, Miriam Rice, Mary Hallesy, Sue Beittel, Jean Starkweather, Mary Hill, Dudley Hubbard, Aubrey Burns, Dr. Martin Griffin, Clerin Zumwalt, Katherine Cuneo, Nello and Philip Kearney – and there are more. Rosamond Day was my guide to the mysteries of marine algae and she has gone over that part of the manuscript, as has Robert Setzer at the California Academy of Sciences.

I thank also Howard Allen, former president of Marin Audubon Society, who helped obtain sponsorship of Marin Audubon Society for this project, and William S. Picher, another former president of Marin Audubon Society and now treasurer of Audubon Canyon Ranch.,

Still others have helped with *The Leaf Book*: Phyllis Ellman and Carole Ericson at Old St. Hilary's in Tiburon; Phyllis Lindley and other members of Golden Gate Audubon Society; Peggy Brown and John Kipping at Strybing Arboretum in San Francisco; Doris Leonard and George Collins of Conservation Associates, and through them, Dr. Hans Jenny of the University of California at Berkeley.

My friends Stephen and Lucienne Dimitroff, Arlen (who suggested the title) and Clare Philpott, Susan Bayerd, Beverly Savitt, and also Rembert Kingsley, Kay Corlett, and Bonnie Margolis, have all helped specifically with the art and technique of plant printing or with the practicalities of book publishing.

I am grateful to my son, Jonathan Perry, for his interest in my progress, and to my mother and my family for their never-failing encouragement.

Finally I owe special thanks, very special thanks, to my friends Margaret and Alfred Azevedo. It was their most practical help that made it possible for me to complete work on *The Leaf Book*.

IDA GEARY

Introduction

The Leaf Book is a field guide to "leaf through" when identifying Northern California native plants. It is the kind of book I wanted when I first became interested in knowing the names of plants: you hold the flower or plant in one hand, and with the other you "leaf through" the book until you find the picture of your plant – and its name.

All 360 pictures in *The Leaf Book*, with a few exceptions, are plant prints, made from the plants themselves and reproduced in life-size, one to a page. They are arranged in family order within seven sections:

MARINE ALGAE

FUNGI, LICHENS AND MOSSES

FERNS AND FERN ALLIES

GRASSES, SEDGES AND RUSHES

WILDFLOWERS

SHRUBS

TREES

A short introduction precedes each section and the major plant families are described briefly as they occur.

Each plant picture is captioned with its common name, scientific name, flower color, if it has a flower, size, or where or when it may be found – all in non-technical language and always with the beginner in mind.

Within each of the seven sections the plants are arranged in family order from the oldest and most primitive plants first to the plants called advanced, those which came later in the evolutionary scale. For the order I have followed John Thomas Howell's *Marin Flora*, except for marine algae, fungi, lichens and mosses, which are not included in that book but which Mr. Howell suggested I include in this.

The family order brings related plants together and this helps the beginner toward more intelligent identification. Families are based generally on characteristics of flowers and fruits, because these change less readily under differing soil and climatic conditions. But leaves also are unique and provide many important and helpful characters for identification purposes, and hence the emphasis on leaves in *The Leaf Book*.

Marin algae, fungi, ferns, and grasses are natural categories, and as more primitive plants they are arranged first in the book. Wildflowers, shrubs, and trees, however, are arbitrary categories, but since I notice first whether a plant is a tree, a shrub or a wildflower perhaps other beginners will appreciate this arrangement.

Most of the plants in *The Leaf Book* were collected in Marin County, north of San Francisco, a crossroads for northern and southern California plants and rich in plant life. I also collected in Sonoma County and Mendocino County in the Coast Ranges, and in the Sierra Nevada near Yosemite National Park and near Reno. Therefore the book should be useful for an extended range across northern California. Occasionally I used young leaves for prints because of their smaller size and these may not always be as characteristic as mature leaves – but they are in the minority.

I have also not been consistent as to native and introduced plants. I have omitted the large and well-known introductions – ornamental trees and shrubs – but I have included some smaller introductions – grasses, filaree, poison hemlock – because they are so widespread.

Conversely, some common plants are not included because they defied printing and some, for one reason or another, I was unable to collect. But then it has been my experience that no one book ever contains all the information the reader wants.

John Muir said that you have to know a place to want to save it. And Aldo Leopold said you can be ethical only in relation to something you can see, feel, understand, love – in other words, know. Know, then, these flowers of the fields – and here John Thomas Howell has added to my introduction – "and help save them!"

Marine Algae

Marine algae, or seaweeds, are primitive plants that do not flower but reproduce by means of spores. They are classified by color – Green, Brown, and Red Algae – but the colors are often confusing, the Red Algae sometimes looking brownish or greenish, the Green Algae brownish.

Like land plants, seaweeds are diversified and seasonal. Instead of a root, however, they have a holdfast, usually attached to a rock. Instead of a stem they have a stipe; and instead of leaves they have blades.

Marine algae are usually found in one of four zones on rocky beaches: Zone I is high tide or splash zone; Zone II is sometimes covered with water; Zone III is practically always covered with water; and Zone IV, below low tide, is never uncovered at all.

Low tide is the best time to explore rocky beaches for marine algae.

Some marine algae are represented here in plant prints, others by pressings.

GREEN ALGAE *chlorophyta*

SEA LETTUCE
ulva sp.

BROWN ALGAE *Phæophyta*

HORSETAIL
Heterochordaria abietina

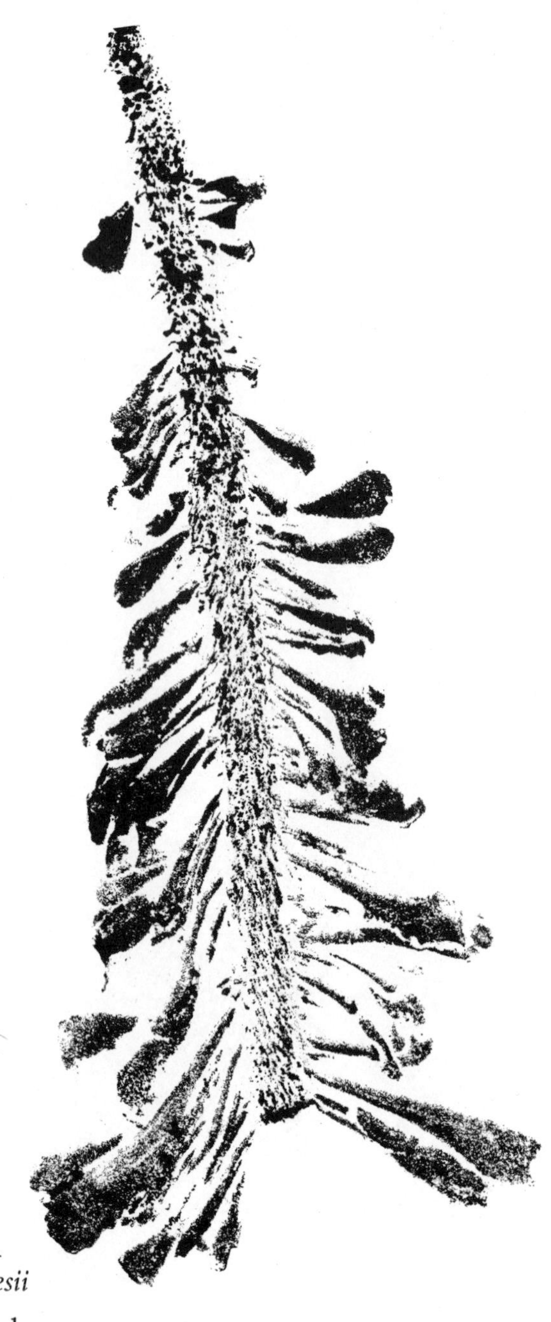

Feather Boa
Egregia menziesii

Can be 25 feet long

Cystoseira osmundacea

Can be 20 feet long

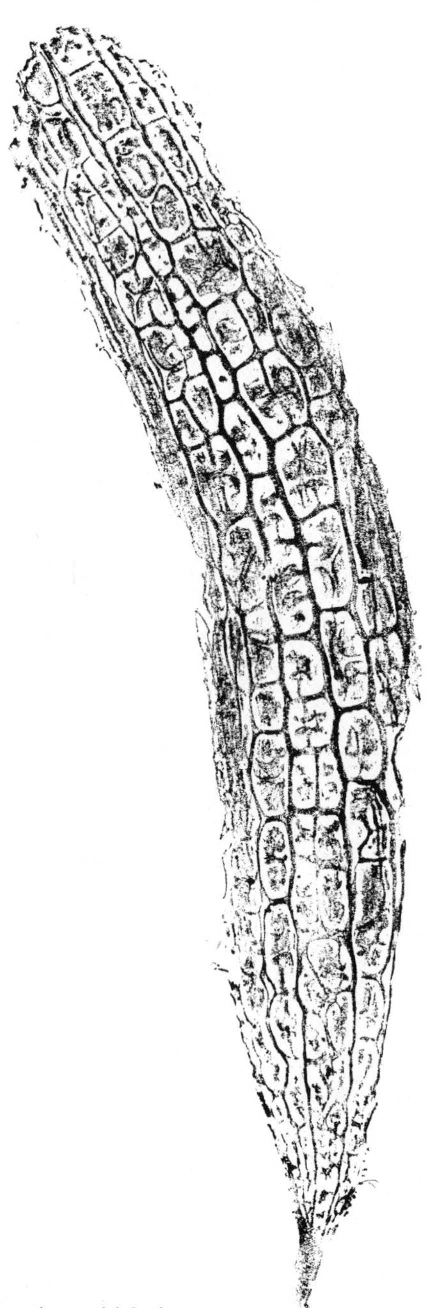

costaria costata

Prominent ribs in broad blade

Sea Palm
postelsia palmæformis

On farthest out rocks

Pelvetiopsis limitata

High on rocks; shown fruiting

ROCKWEED
Fucus distichus

High on rocks; shown fruiting; prominent mid-rib

RED ALGAE *rhodophyta*

cryptosiphonia woodii
(a pressing)

Plocamium pacificum
(a pressing)

Grows on other plants

Gigartina agardhii

Has ridge on edges

Turkish Towel
Gigartina papillata

Callithamnion pikeanum

SEA LACE
Microcladia coulteri
(a pressing)

Grows on other plants

POLYSIPHONIA
(a pressing)

Rhodomela larix

Looks like a brown mat on rocks offshore

Ostric-plumed Hydroid

Not an algae but a community of animals

Fungi, Lichens and Mosses

FUNGI, without flowers or seeds, propagate by spores. They also lack chlorophyll and are either parasitic or saprophytic.

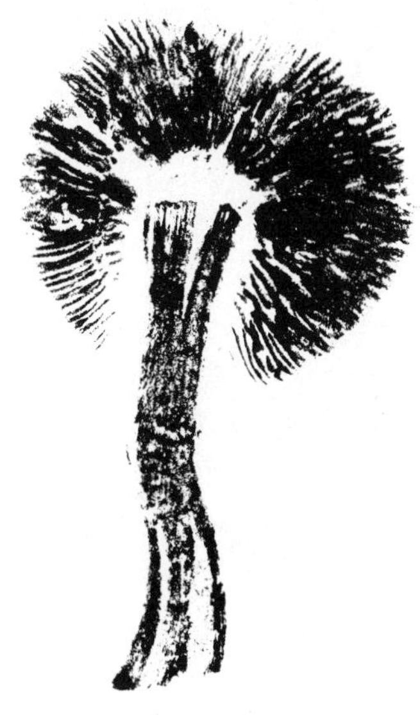

MUSHROOM (a plant print)
Agaricus sp.

MUSHROOM (spore prints)
Agaricus spp.

Spore prints are made by removing the stem from the fungus, placing the gill or spore-bearing side down over paper, and covering it with a glass bowl. After a few hours the spores will have dropped on the paper. The print should be sprayed immediately with a fixative.

LICHENS are dual organisms, plants composed of a fungus and an alga. The fungus is believed to be mildly parasitic on the alga.

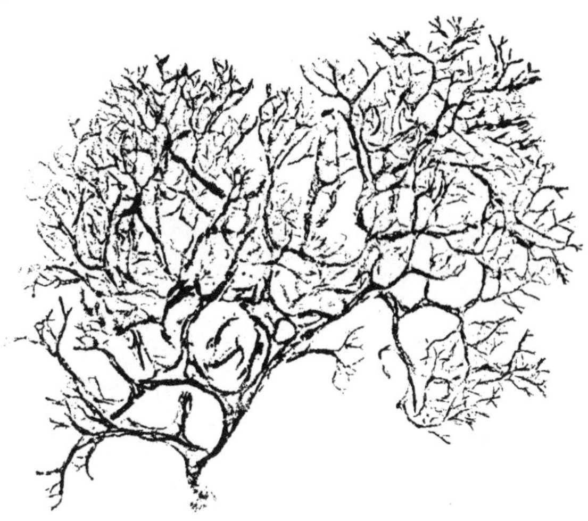

STAGHORN; WOLF LICHEN
Letharia vulpina

Bright yellow-green, on bark of trees

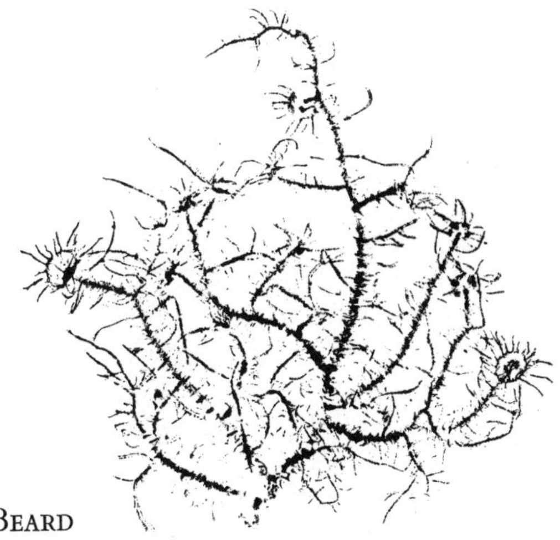

OLD MAN'S BEARD
Usnea sp.

A lichen, shown here with fruiting bodies

Mosses, usually just a few inches high, grow from spores produced in capsules.

Mnium menziesii

In soil; along Pacific Coast

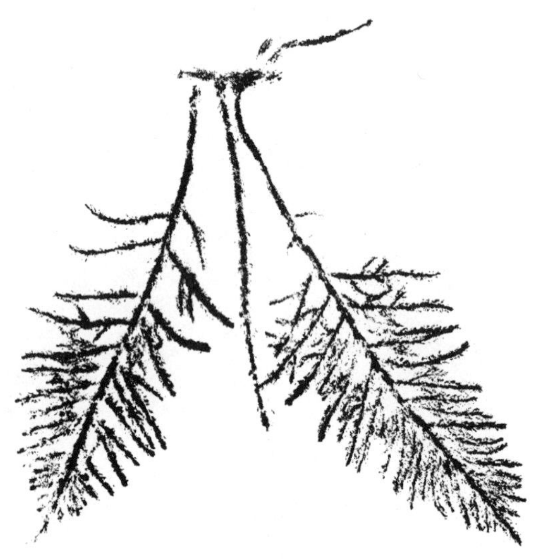

Dendrolasia abietina

On bark of trees

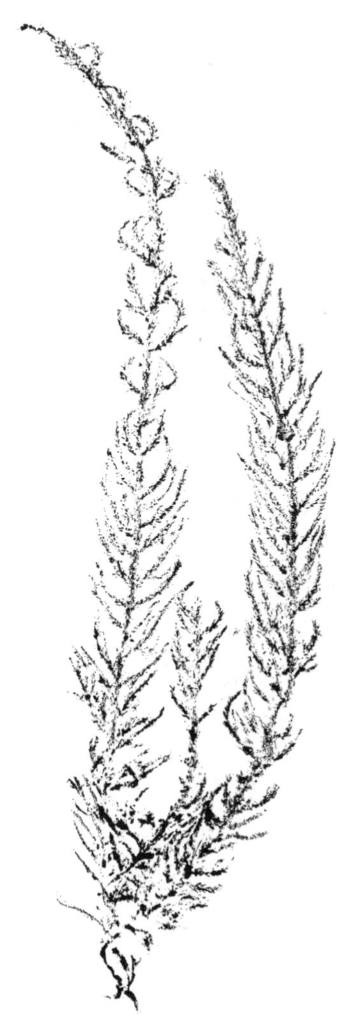

FEATHER MOSS
Isothecium sp.

NOTES

Ferns and Fern Allies

FERNS (*Polypodiaceæ*), unlike most higher plants which produce seed, reproduce from spores. Spores develop on the undersides of the fern leaflets, drop into the ground, and there develop into a prothallus, or gametophyte plant. This is a tiny, flat, heart-shaped plant, the size of a pencil eraser.

The gametophyte plant, in its turn, produces male and female sex organs, microscopic in size, on its underside. They, after fertilization, produce the conspicuous leafy plants we know as ferns. This life cycle process, involving two generations, is called the "alternation of generations."

HORSETAILS (*Equisetaceæ*), primitive plants with hollow, jointed stems, are related to the ferns and also reproduce from spores.

MOSS FERNS (*Selaginellaceæ*), low creeping plants with solid stems, are also called Spike Mosses. They too reproduce from spores.

FERN FAMILY *Polypodiaceæ*

BRITTLE FERN
Cystopteris fragilis

Small and fragile

Lady Fern
Athyrium filix-femina

Tall, fragile, lower leaflets narrow

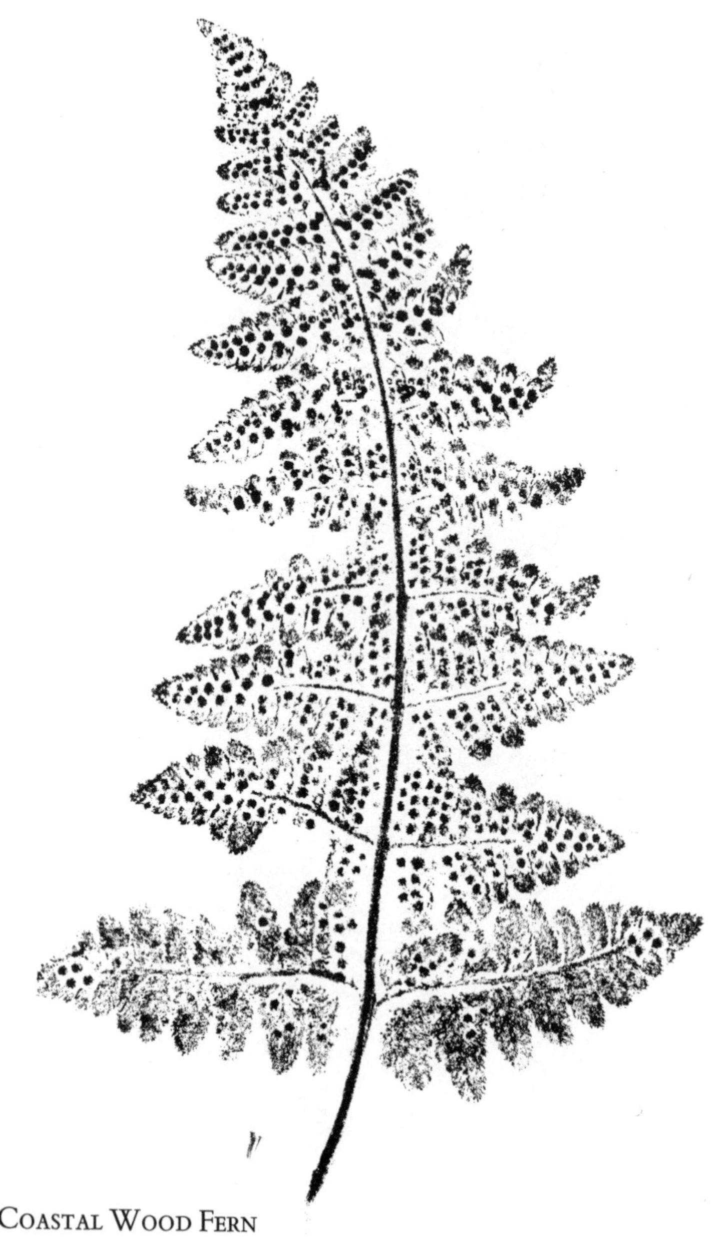

Coastal Wood Fern
Dryopteris arguta

Sturdy, lower leaflets broad

Spreading Wood Fern
Dryopteris austriaca

Leaf 3 times pinnate

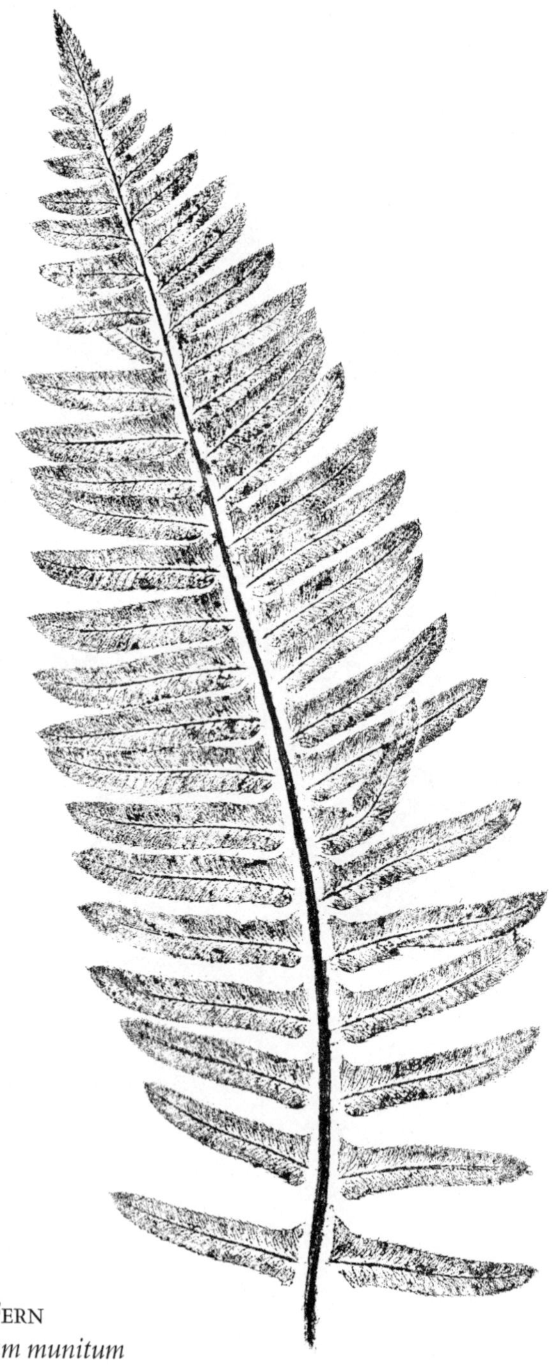

Sword Fern
polystichum munitum

"Hilts" on leaflets; large clumps

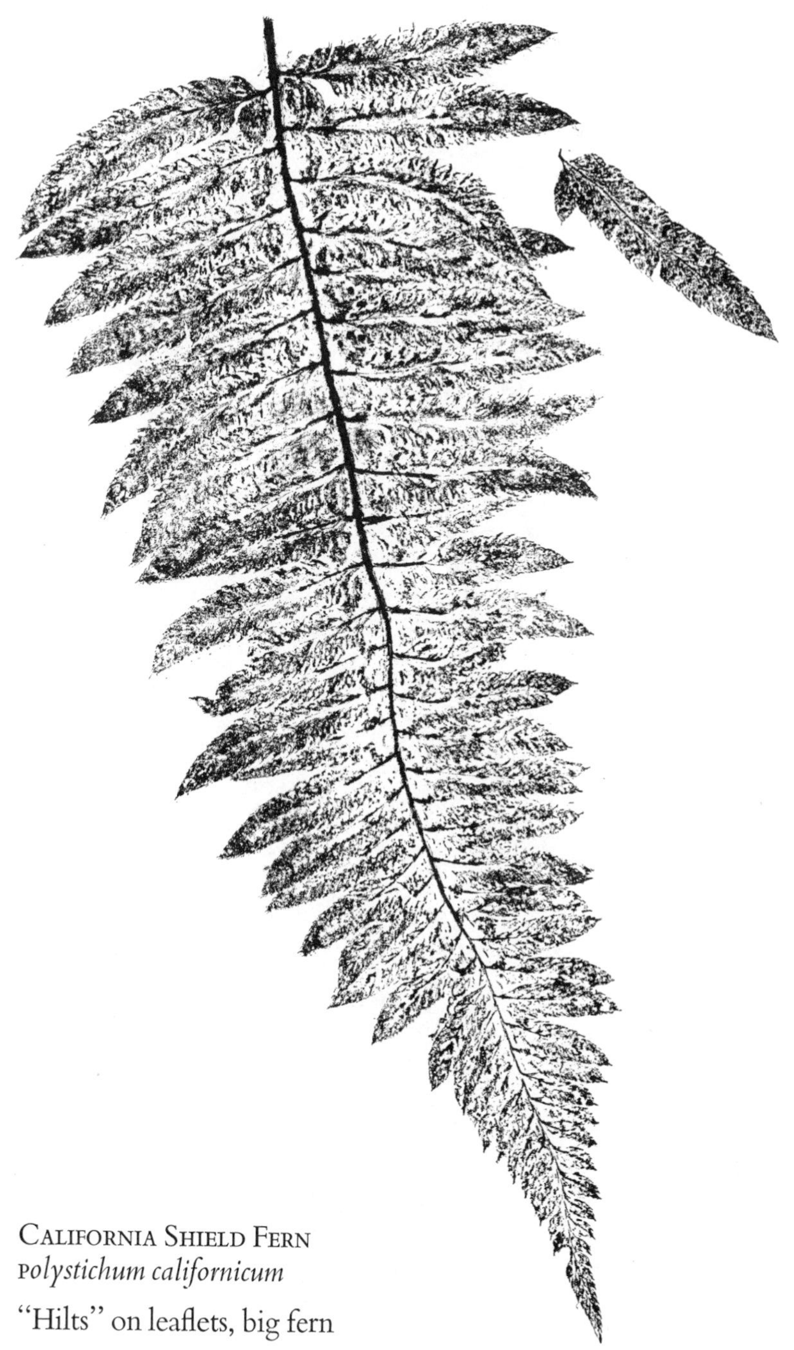

California Shield Fern
polystichum californicum

"Hilts" on leaflets, big fern

Dudley Shield Fern
Polystichum dudleyi

"Hilts" on leaflets, lacier than other shield ferns; rarer

Chain Fern
woodwardia fimbriata

Tall, in wet places; sori in "chains"

Leather Fern
polypodium scouleri

Low and "leathery" on rocks and tree trunks

California Polypody
Polypodium californicum

Leaflets have rounded ends, size and shape variable, common, on rocks

LICORICE FERN
polypodium glycyrrhiza

Leaflets have pointed ends; occasional, on rocks and logs

GOLDBACK FERN
Pityrogramma triangularis

Dry banks; golden undersides

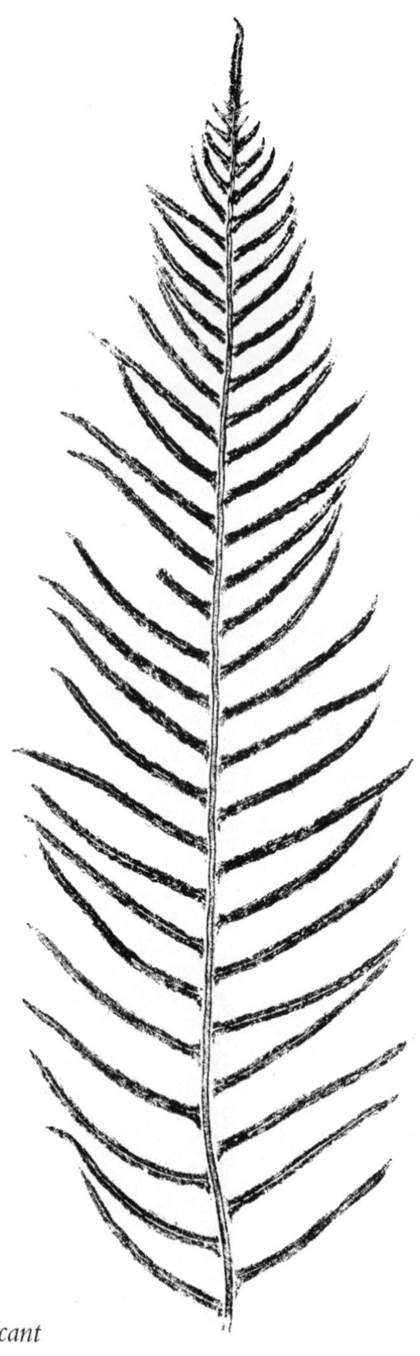

DEER FERN
Blechnum spicant

Two kinds of leaves, this one tall, thin and fertile

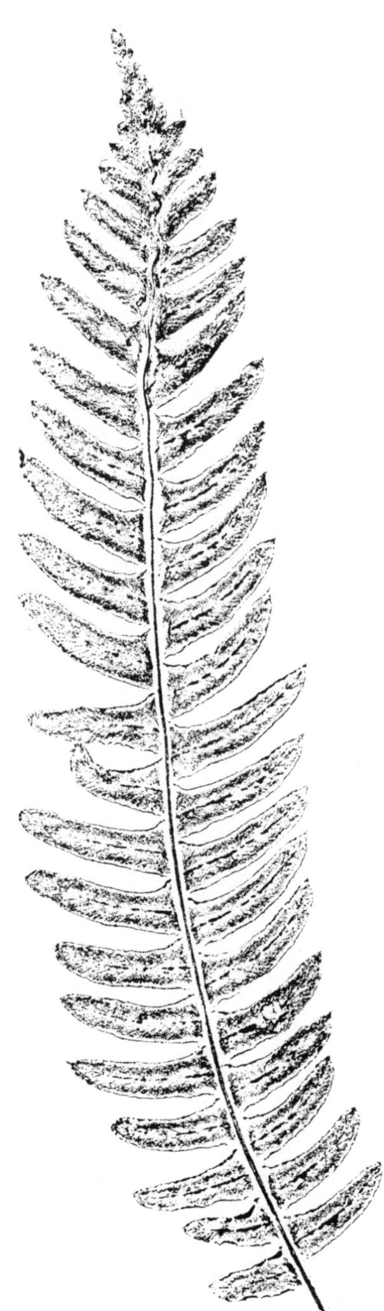

DEER FERN
Blechnum spicant

Sterile leaf, shorter than fertile leaf

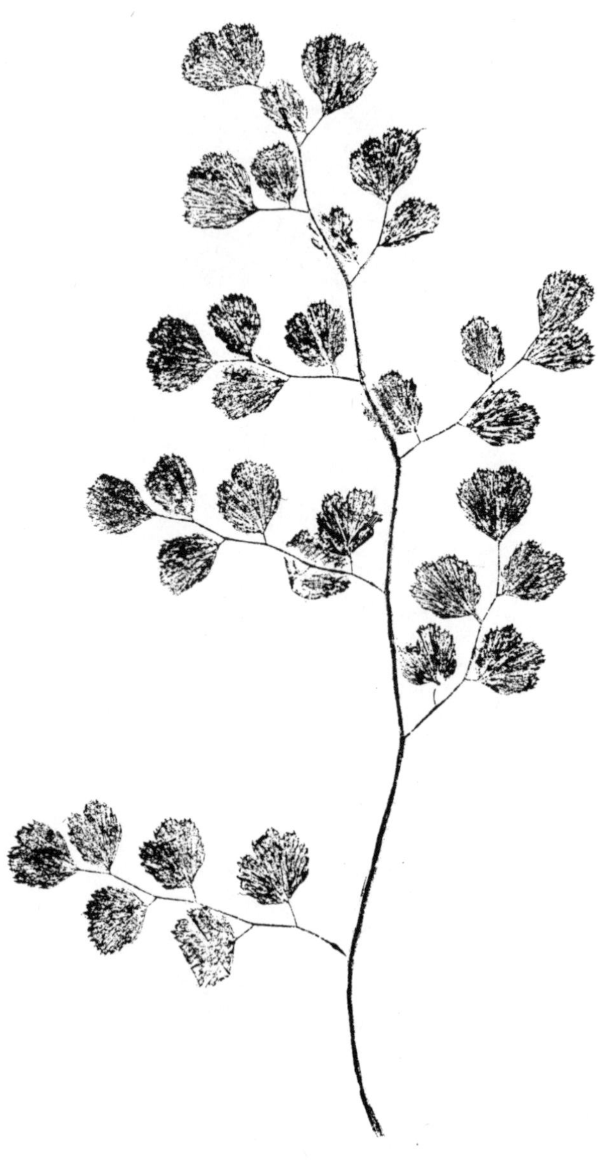

Maidenhair Fern
Adiantum jordani

WESTERN FIVEFINGER FERN
Adiantum pedatum

Can have 3 to 8 "fingers"

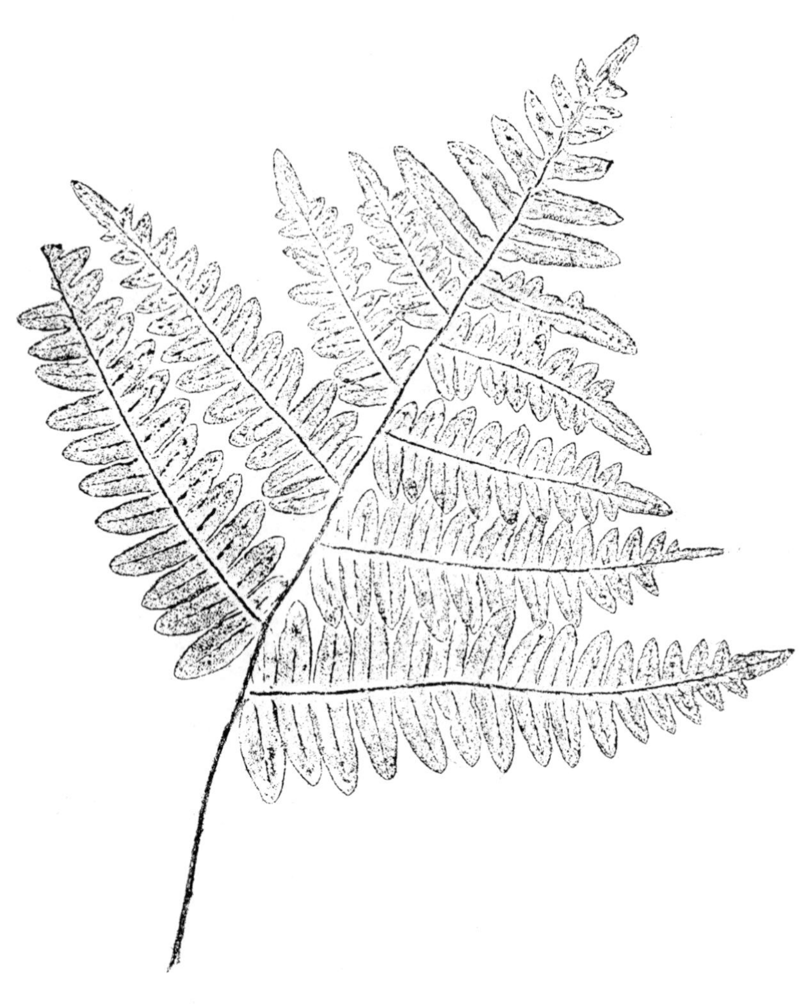

Western Bracken Fern
Pteridium aquilinum

Common, in sunny places

California Lace Fern
cheilanthes californica

Rocks and cliffs in Coast Ranges; also Sierra Nevada

Serpentine Fern
Cheilanthes siliquosa

Clustered at base of serpentine rocks

Lace Fern
Cheilanthes gracillima

In crevices on sun-warmed rocks

Coville's Lip Fern
cheilanthes covillei

4 to 12 inches long, in rocky crevices

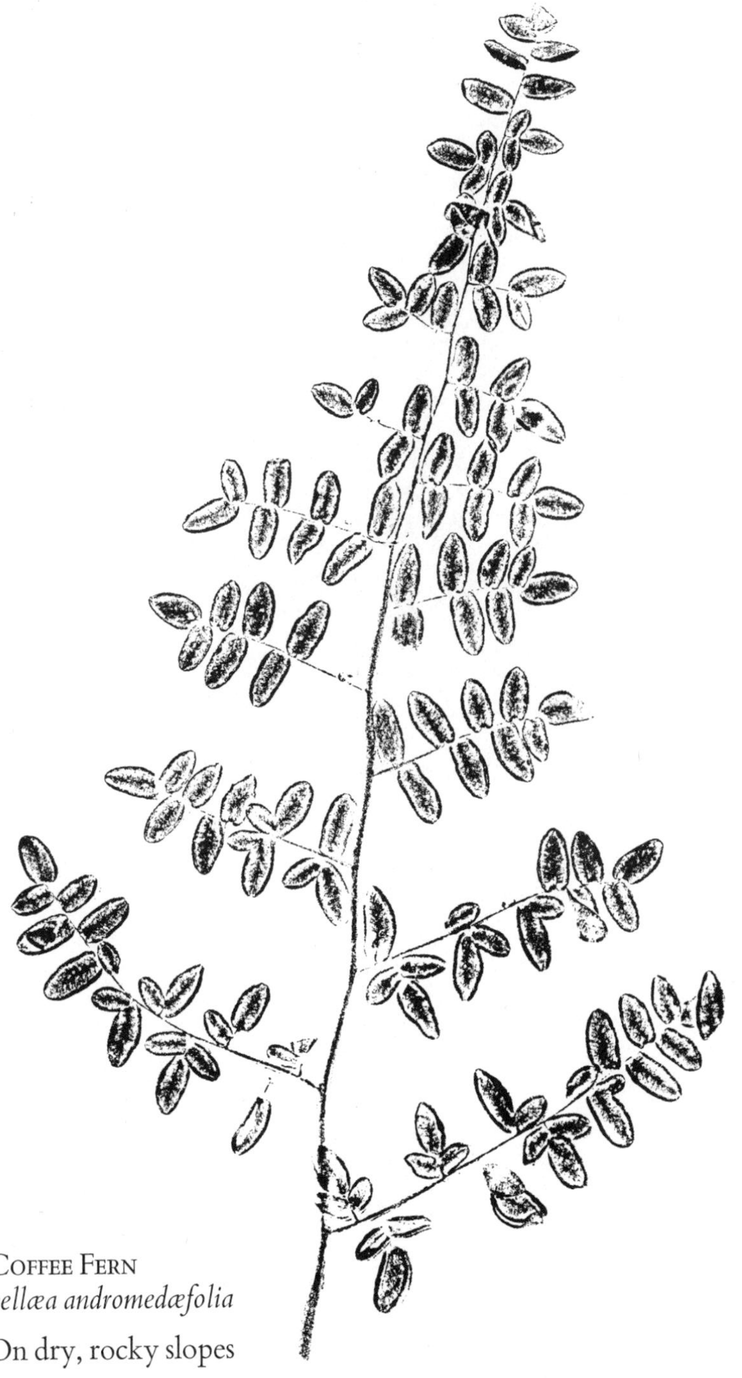

COFFEE FERN
Pellæa andromedæfolia

On dry, rocky slopes

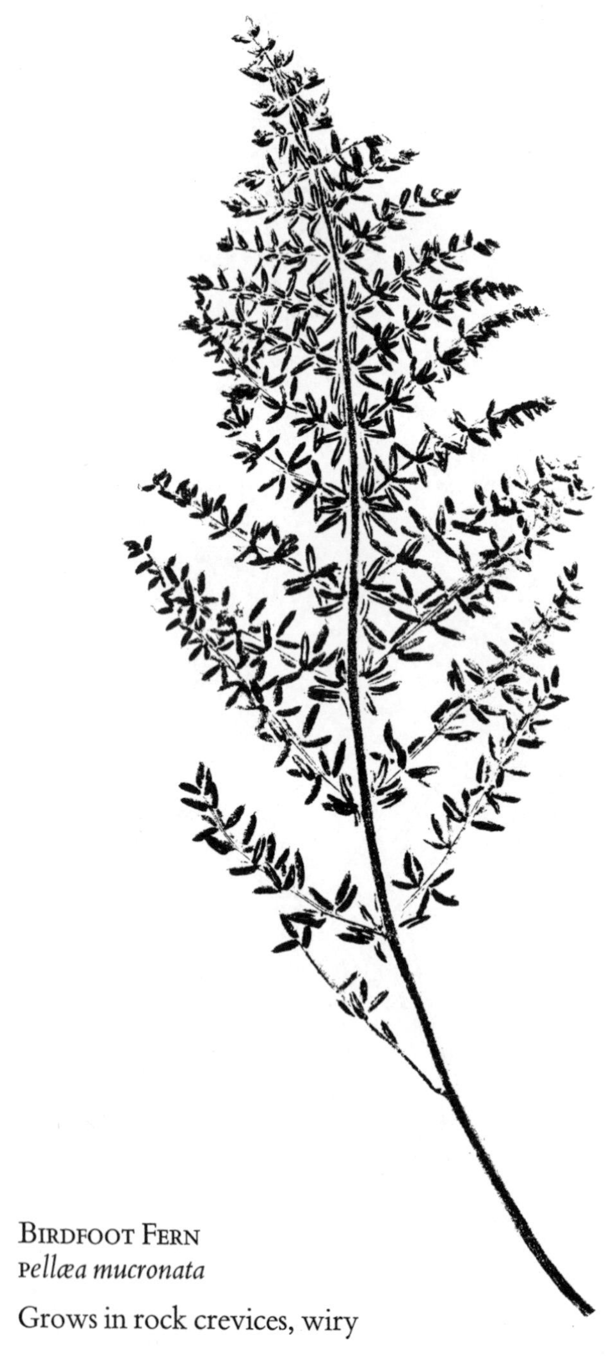

Birdfoot Fern
Pellæa mucronata

Grows in rock crevices, wiry

HORSETAIL FAMILY *Equisetaceæ*

GIANT HORSETAIL
Equisetum telmateia

Fruiting spike, *right*, taller than sterile stem

MOSS FERN FAMILY *selaginellaceæ*

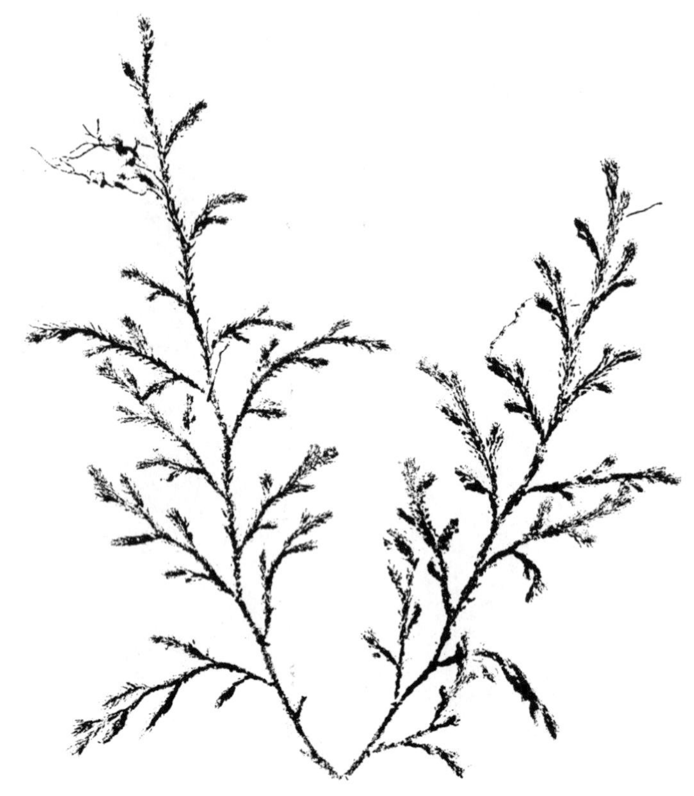

Wallace Selaginella
selaginella wallacei

On rocks in coast redwood forests

CAT-TAIL FAMILY *Typhaceæ*

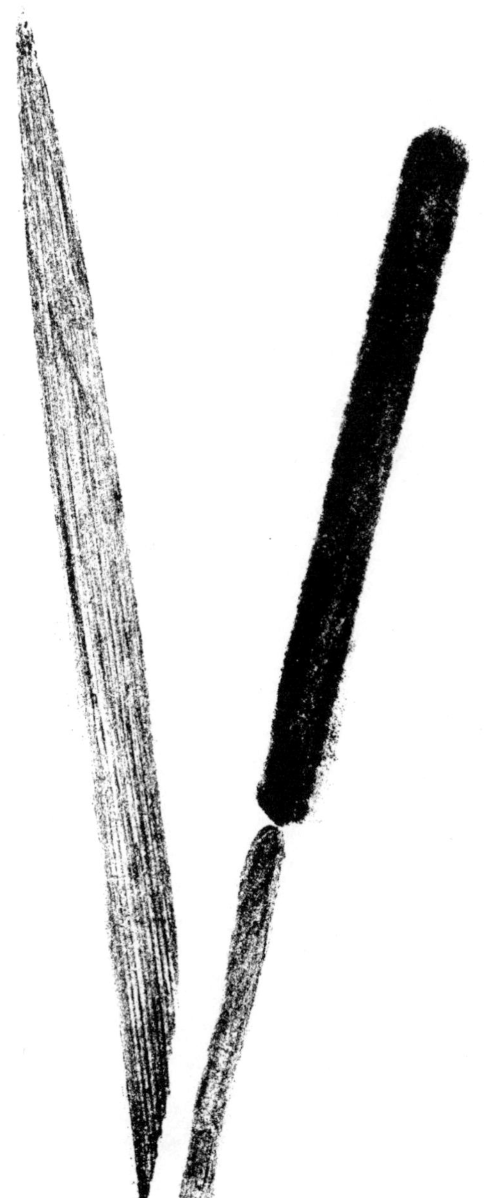

CAT-TAIL
Typha latifolia

Marshy places, 3 to 6 feet high, dark brown fruiting spike and long flat leaves

BUR-REED FAMILY *sparganiaceæ*

BUR-REED
sparganium greenei
Tall plant in freshwater marshes

Grasses, Sedges and Rushes

GRASSES, SEDGES AND RUSHES are classified as monocotyledons, or monocots, flowering plants with one seed leaf or cotyledon, parallel-veined leaves, and flower parts in threes. In addition to grasses, monocots include the Lily, Iris, and Orchid families. Monocots are usually herbs, although some are trees.

Members of the Grass Family (*Gramineæ*) have hollow stems, usually round, and swollen nodes. They are low-growing, widespread, familiar looking, and difficult for the beginner to identify. A hand lens or microscope reveals that grass spikelets are composed of tiny flowers, or florets, which have no petals. Other parts of the spikelets are glumes, the lemmas and paleas, and awns. The fruit is called a grain.

An excellent key to the Grass Family is contained in John Thomas Howell's *Marin Flora*.

The Sedge Family (*Cyperaceæ*) is characterized by usually solid stems, without nodes, and the stems can be triangular, four-sided, round, or flat. Like grasses, sedges also have florets arranged in spikelets. Tule, bulrush, and papyrus are all sedges, which can grow in either wet or dry soil. The fruit is called an achene.

Members of the Rush Family (*Juncaceæ*) resemble grasses and sedges but the tiny flowers under a lens or microscope look more like a miniature lily. Rushes grow mostly in wet places and the fruit is called a capsule.

EELGRASS FAMILY *zosteraceæ*

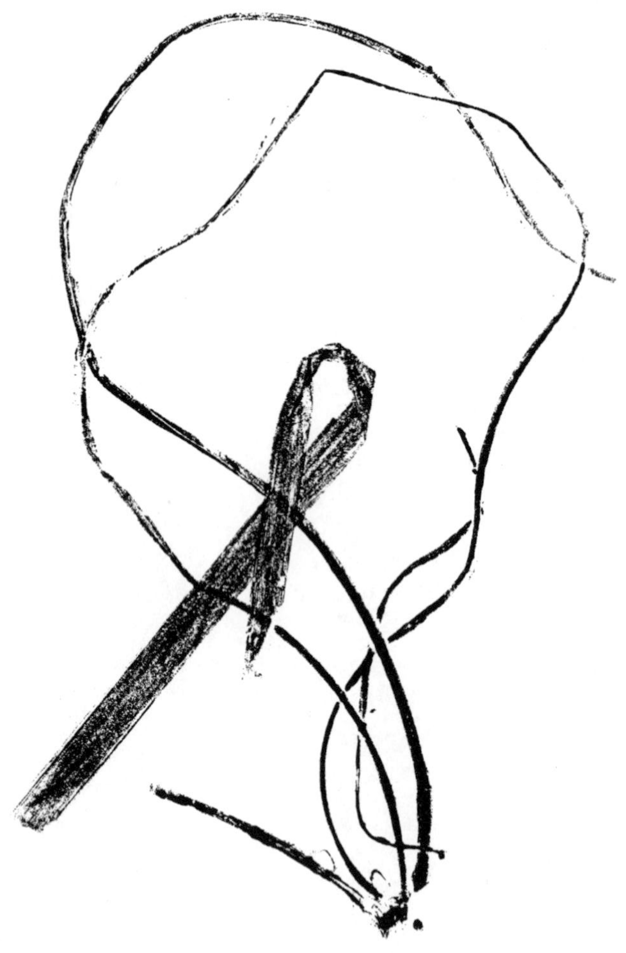

EELGRASS, *zostera marina*

Broad-leafed grass growing on bottom of shallow bays near ocean

SURFGRASS, *phyllospadix sp.*

Narrower leaf than Eelgrass, grows on rocks or on reefs of ocean shores (*combined print*)

GRASS FAMILY *Gramineæ*

DOWNY CHESS [FESCUE TRIBE]
Bromus tectorum

Introduced

FESTUCA
Festuca rubra

Coast and inland

Douglas Blue Grass
Poa douglasii

Coastal dunes

Rattlesnake Grass
Briza maxima

Introduced

LITTLE RATTLESNAKE GRASS
Briza minor

Introduced

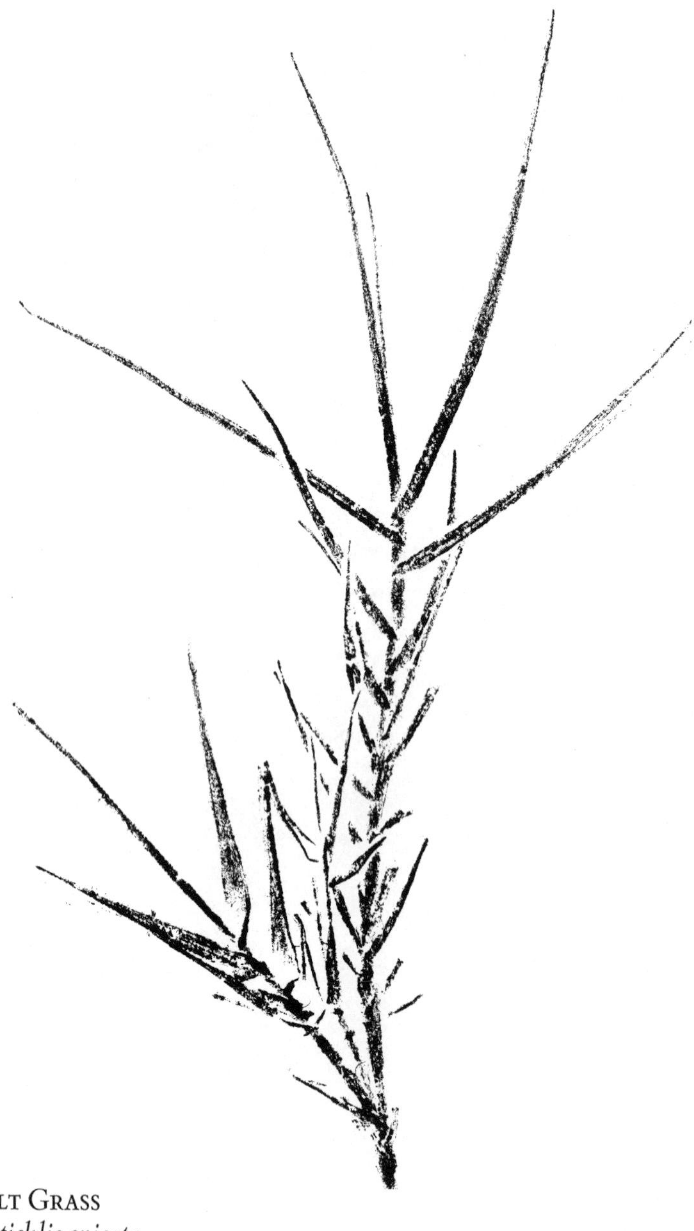

Salt Grass
Distichlis spicata

Salt marshes

DOGTAIL GRASS
Cynosurus echinatus

Introduced

WILD RYE
Elymus jepsonii

AMERICAN DUNE GRASS
Elymus mollis

Squirreltail Grass
sitanion jubatum

[OAT TRIBE]

HAIR GRASS
Deschampsia holciformis

Shiver Grass
Aira caryophyllea

5-inch-high reddish furze (Introduced)

WILD OATS
Avena fatua

Introduced

Velvet Grass
Holcus lanatus

Violet-tinted

[TIMOTHY TRIBE]

Timothy Grass
Calamagrostis ophitidis

NEEDLE GRASS
stipa pulchra

[CHLORIS TRIBE]

PACIFIC CORD GRASS
spartina foliosa

Pioneer in salt marshes, it grows in the water; a tall grass

California Vanilla Grass
Hierochloe occidentalis

In redwood forests

SEDGE FAMILY *cyperaceæ*

scirpus cernuus

4 inches tall, in wet places

Bulrush; Tule
Scirpus validus

5 feet tall, in ponds and marshes

RUSH FAMILY *Juncaceæ*

SALT RUSH
Juncus leseurii

Salt marshes and dunes, 1–3 feet high

Toad Rush
Juncus bufonius

4 inches tall, in wet places

Wildflowers

Wildflowers, native perennial or annual herbs, are angiosperms divided into two categories: the monocotyledons, or monocots, and the dicotyledons, or dicots.

The monocots, which some botanists believe are the older and more primitive group, have flower parts in threes and simple leaves which are parallel-veined. Grasses and the Lily, Iris and Orchid families are classified as monocots.

The dicots, flowering plants with two seed leaves, or dicotyledons, usually have flower parts occurring in fours or fives, and net-veined leaves. This is the much larger group, including also shrubs and trees.

Monocotyledons

LILY FAMILY *Liliaceæ*

Flower parts in 3's, leaves parallel-veined; largest family of flowering plants

MARSH ZIGADENE
Zigadenus fontanus

In wet places, often on serpentine; cream-colored flower

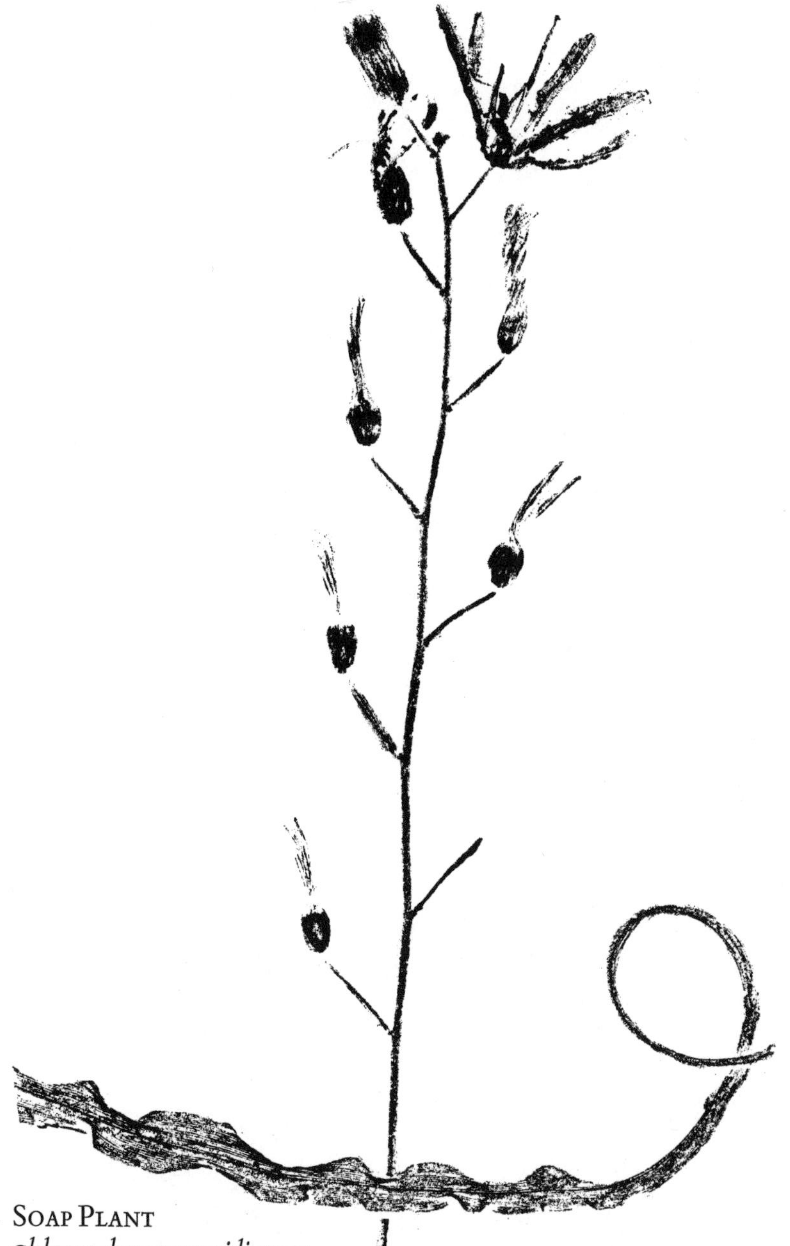

Soap Plant
Chlorogalum pomeridianum

White star-like flowers open late in afternoon; rippled leaves

Wild Onion
Allium lacunosum
White flower, open places

COMMON BRODIAEA; BLUE DICKS
Brodiæa pulchella

Purple flower; all over; long narrow basal leaves of Brodiaeas are few and obscure

Ithuriel's Spear
Brodiæa laxa

Loose purple flower; grassland

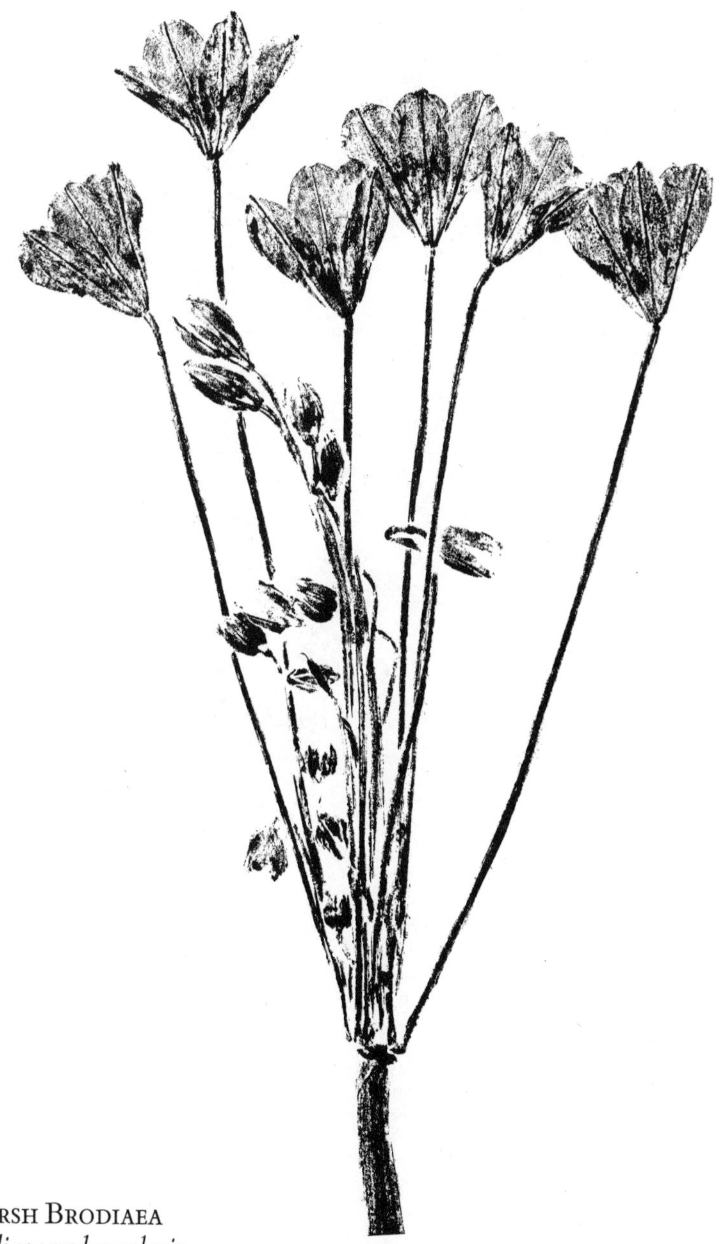

MARSH BRODIAEA
Brodiaea peduncularis

White flower with purple midvein; wet places

Bronze Bells; Fritillary
Fritillaria lanceolata

Bronze-purple mottled flowers; on banks in woods in filtered sunlight

Pussy-Ears
Calochortus tolmiei

Lavender flower; grassy places near coast

Yellow Mariposa
Calochortus luteus

Sturdy summer lily of dry, open places; at first glance looks like California poppy

FETID ADDER'S TONGUE
scoliopus bigelovii

Brownish flower, brown-spotted leaves; fetid refers to odor—like old, wet dog. Blooms early

TRILLIUM
Trillium ovatum

White flower, on wet mossy banks; blooms early

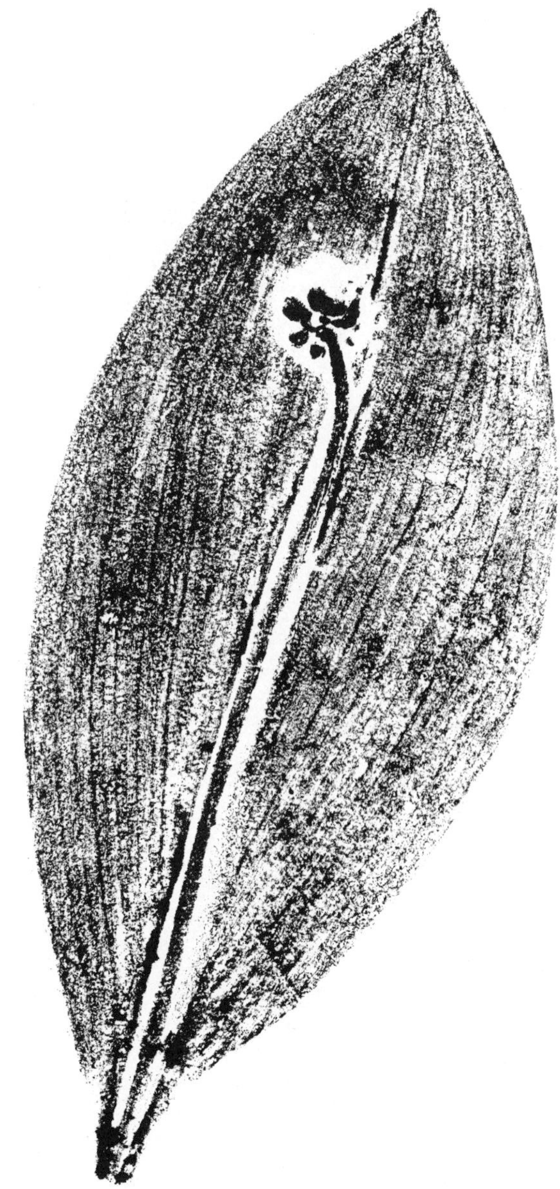

Clintonia andrewsiana

In deep forests; rose flowers; here young flower stalk has not yet grown out of a basal leaf

Fat Solomon's Seal
smilacina amplexicaulis

White flowers in branching clusters, or panicles; if more slender flower cluster in a raceme, without branching, it is Slim Solomon, *smilacina sessilifolia*; shady places

Fairy Bells
Disporum smithii

Creamy flowers almost hidden; shady places

IRIS FAMILY *Iridaceæ*

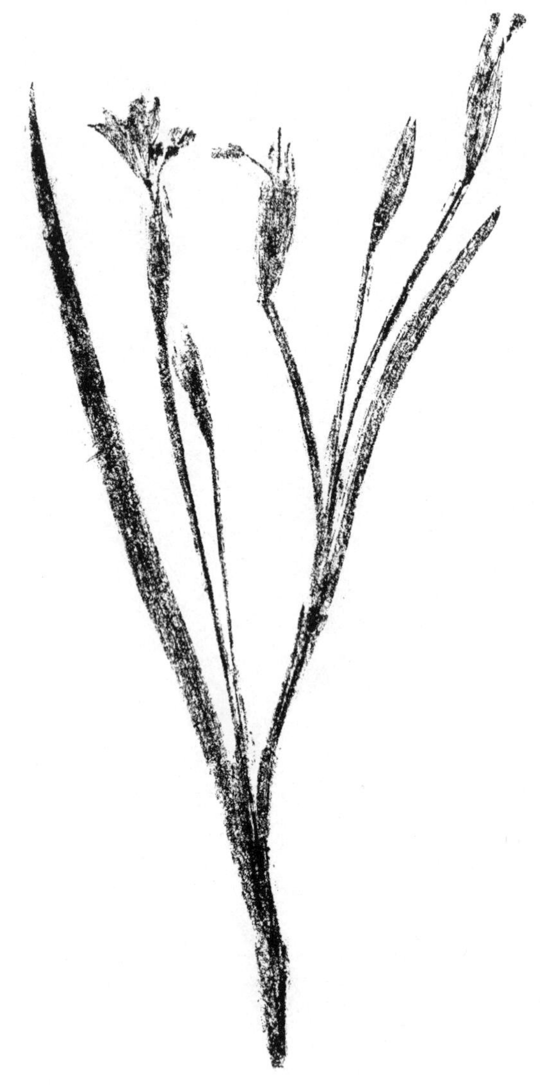

GRASS IRIS; BLUE-EYED GRASS
Sisyrinchium bellum

Purple flower, in grassy places

GROUND IRIS
Iris macrosiphon

Deep bright purple flowers, low, in open fields; fragrant

ORCHID FAMILY *orchidaceæ*
Irregular flowers, one petal forming a "lip." Large, diverse family with 25,000 species.

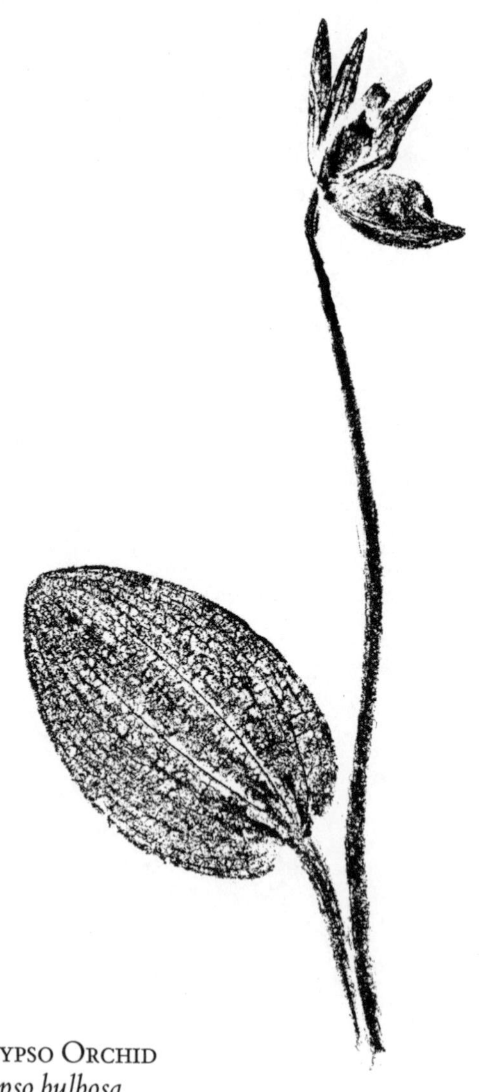

Calypso Orchid
calypso bulbosa

Single pink flower on short stem, single basal leaf; in leaf litter, shady places

REIN-ORCHIS
Habenaria greenei

Slender spikes of white flowers, here in bud, on slopes near coast

Dicotyledons

NETTLE FAMILY *Urticaceæ*

STINGING NETTLE
Urtica californica

Tiny stinging hairs make plant painful to touch. This plant print was possible only because plant was a very young one.

BIRTHWORT FAMILY *Aristolochiaceæ*

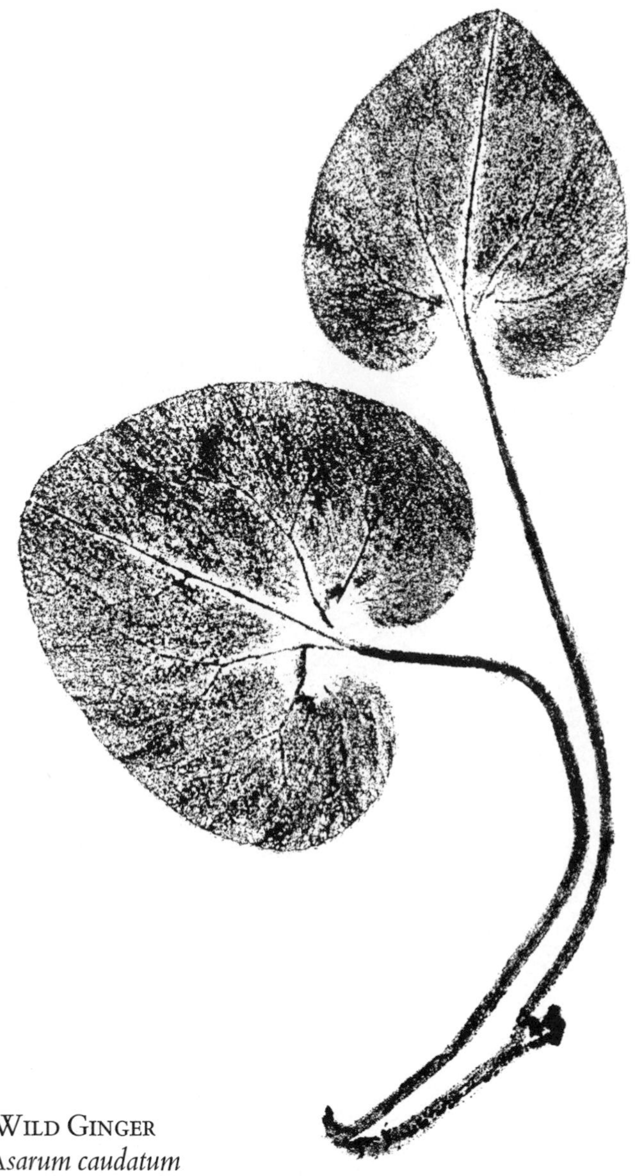

WILD GINGER
Asarum caudatum

Low plant, in redwood forests; brown flower hidden under leaves

HEMP FAMILY *Cannabaceæ*

HEMP; MARIJUANA
Cannabis sativa

Related to nettles but lacks stinging hairs (Introduced)

BUCKWHEAT FAMILY *Polygonaceæ*

Blooms generally in summer or fall

SMARTWEED; KNOTWEED
Polygonum punctatum

White flower on small, low plants, in wet places

KNOTWEED; SMARTWEED
Polygonum paronychia

Pink flower, on seaside cliffs

Dock
Rumex acetosella
Red flowers (Introduced)

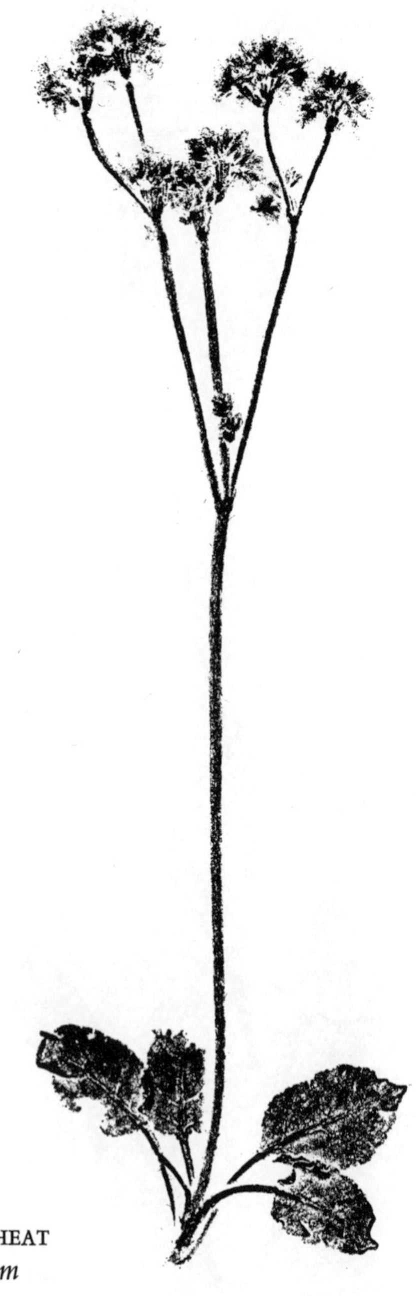

Wild Buckwheat
Eriogonum nudum

Pink flower, on rocky slopes

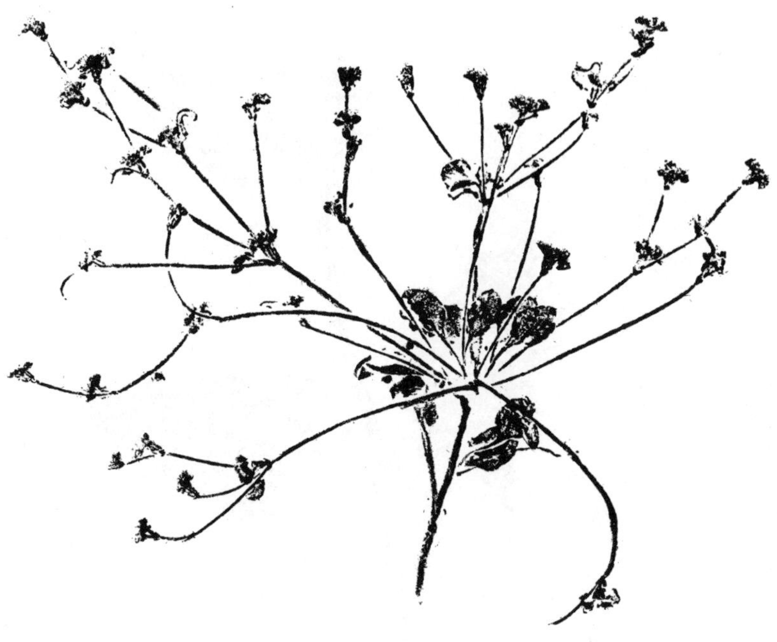

TIBURON BUCKWHEAT
Eriogonum caninum

Tiny pink flower, in dry places

GOOSEFOOT FAMILY *chenopodiaceæ*

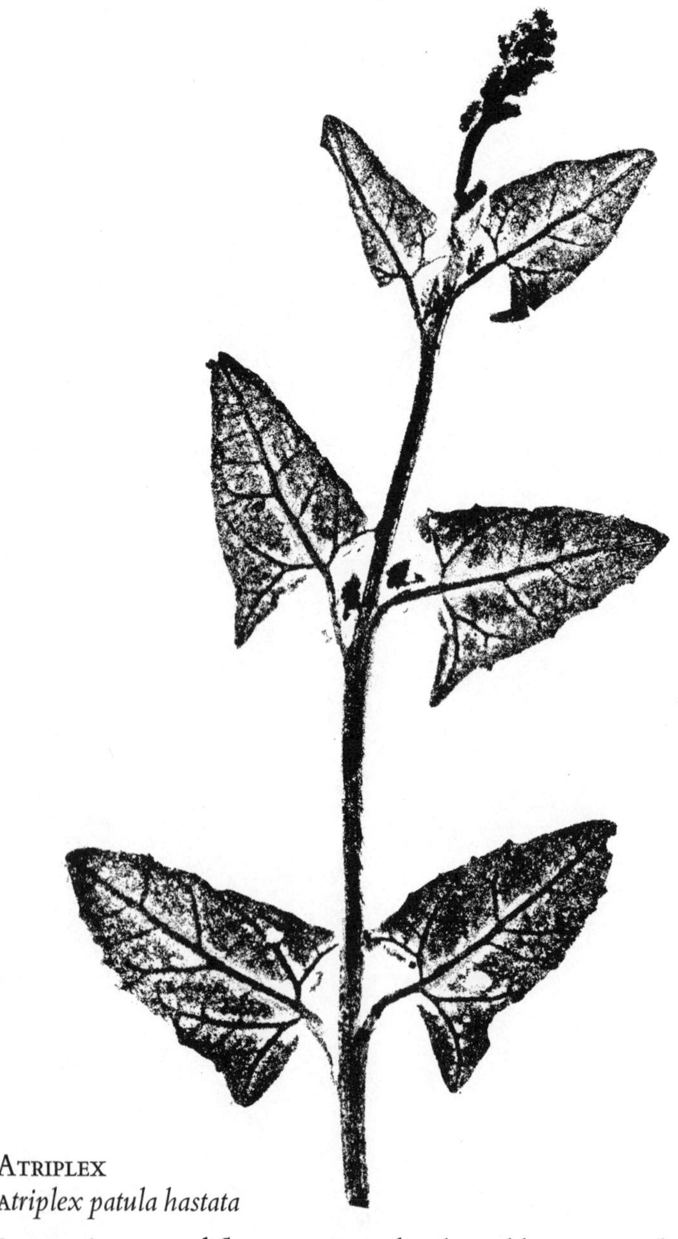

ATRIPLEX
Atriplex patula hastata

Inconspicuous red flower, triangular shaped leaves; in salt marshes

PICKLEWEED
salicornia virginica

Common salt marsh plant

FOUR O'CLOCK FAMILY *Nyctaginaceæ*

SAND-VERBENA
Abronia latifolia

Yellow flowers, succulent leaves, on sand dunes

PURSLANE FAMILY *Portulacaceæ*

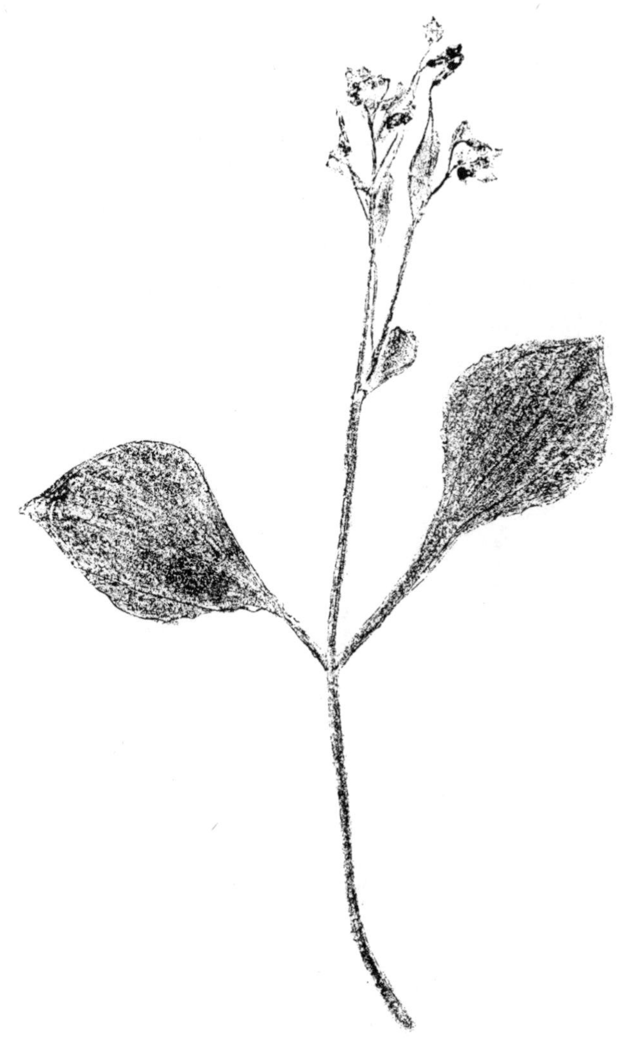

MONTIA
Montia sibirica

Pink flower, in wet soil

MINERS' LETTUCE
Montia perfoliata

Two opposite leaves are fused together; tiny white flower

PINK FAMILY *Caryophyllaceæ*
Opposite leaves, stem swollen at nodes, 5 petals

SANDWORT
Arenaria douglasii

White flower, on dry open slopes

Windmill Pink
silene gallica

Pink flower, near sea coast

INDIAN PINK
silene californica

Bright red flowers, on shaded rocky slopes, late spring to early summer

Sand Spurrey
spergularia macrotheca

Pale flower, on borders of salt marshes

BUTTERCUP FAMILY *Ranunculaceæ*

Buttercups, columbines, and larkspurs; palmately compound leaves; many stamens

BANEBERRY
Actaea arguta

White flowers; red or white berries poisonous

BUTTERCUP
Ranunculus californicus

Forms yellow carpet on grassy meadows

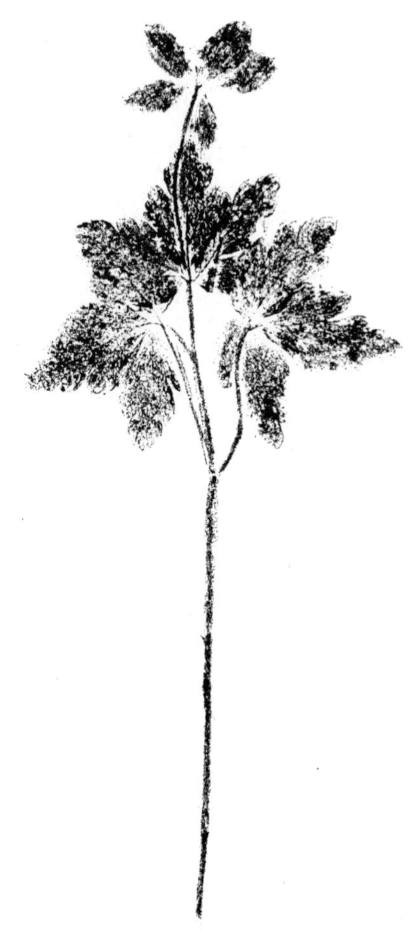

Anemone; Windflower
Anemone quinquefolia

Pale flower, in redwood forests

Meadow Rue
Thalictrum polycarpum

Leafy plant with inconspicuous flowers; male and female flowers on different plants

COLUMBINE
Aquilegia formosa

Yellow flower; on brushy slopes

Blue Larkspur
Delphinium hesperium

Away from the coast

Red Larkspur
Delphinium nudicaule

Often multi-flowered, common on rocky dry hillsides

BARBERRY FAMILY *Berberidaceæ*

Vancouveria planipetala

White flower, in redwoods

POPPY FAMILY *Papaveraceæ*
Sepals drop early, 4 or 6 petals, many stamens

CREAM CUP
Platystemon californicus

Pale yellow flower

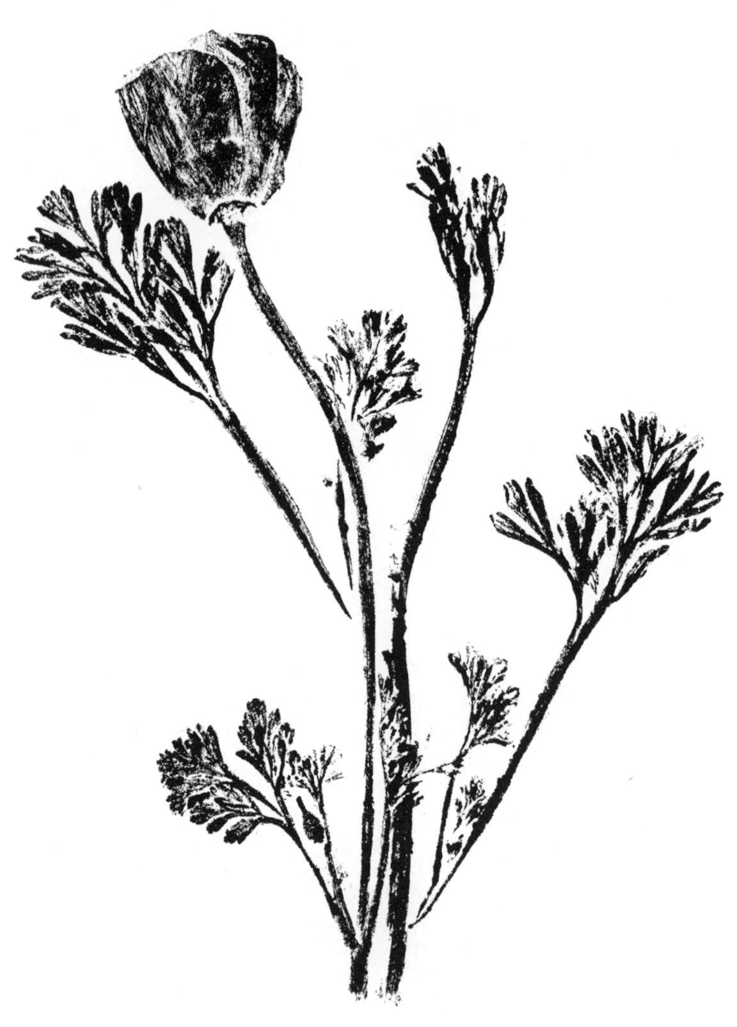

California Poppy
Eschscholzia californica

Orange-yellow flower; California State flower

FUMITORY FAMILY *Fumariaceæ*

BLEEDING HEART
Dicentra formosa

Pink flower, in redwood forests, moist woods and brushy places

MUSTARD FAMILY *cruciferæ*

4 petals in a cross-like pattern; large, world-wide family

BLACK JEWEL FLOWER
streptanthus niger

Garnet flower and reddish spear-like leaf; grows only in John Thomas Howell Botanical Garden, Old St. Hilary's, Tiburon, on serpentine

Jewel Flower
Streptanthus pulchellus

Purple flower is like a tiny jewel on thin, almost invisible stems; dry stony earth, serpentine, Marin County

Milkmaids; Toothwort
Dentaria integrifolia

Pink flower and extremely variable leaves; early spring

WALLFLOWER
Erysimum concinnum

Pale yellow flower; a maritime plant

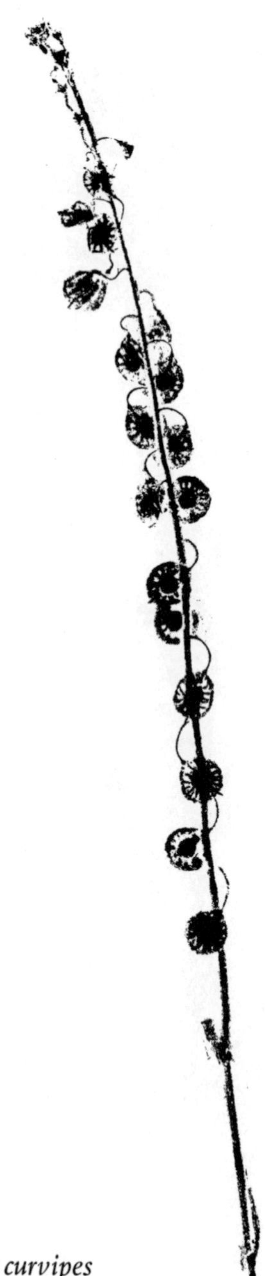

Fringe Pod
Thysanocarpus curvipes

Pods more conspicuous than tiny white flowers; in open places

Shepherd's Purse
capsella bursa-pastoris

Common weed (Introduced)

STONE-CROP FAMILY *crassulaceæ*
Succulent leaves

HENS-AND-CHICKENS
sedum radiatum

Yellow flower springs up from basal rosette of leaves; this is a composite print

SAXIFRAGE FAMILY *saxifragaceæ*
Leaves mostly basal, small flowers mostly 5 petals

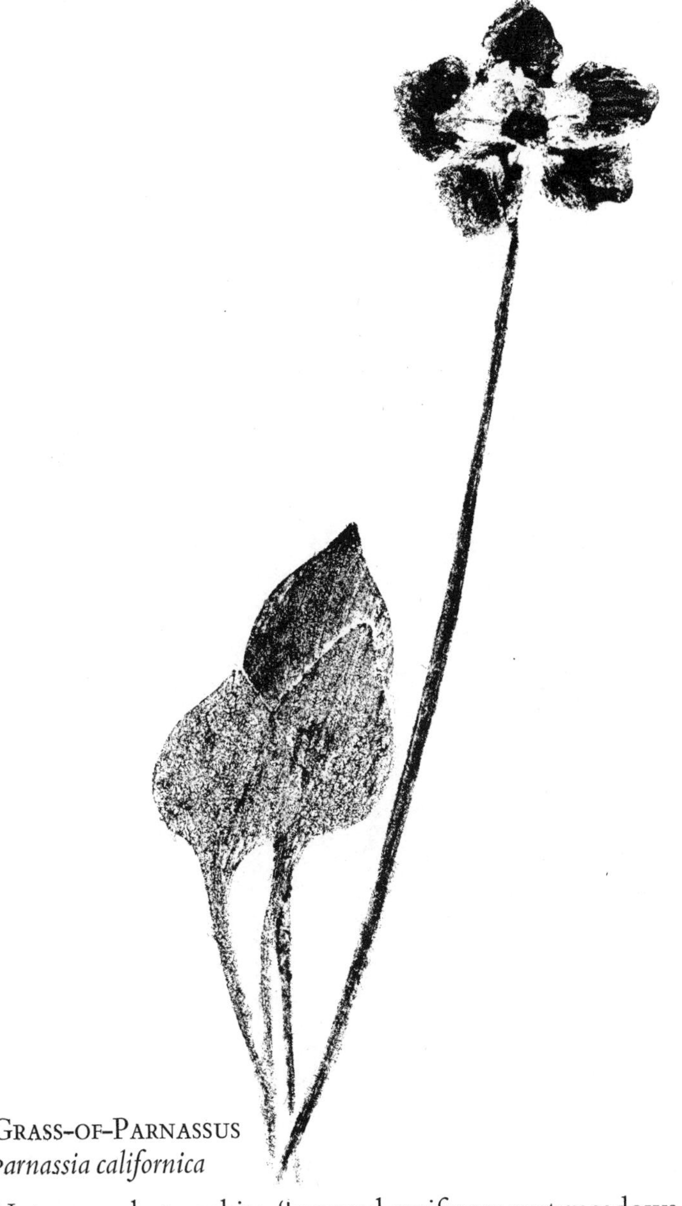

GRASS-OF-PARNASSUS
Parnassia californica

Not a grass but a white-flowered saxifrage; wet meadows, late summer

Sugar Scoop
Tiarella unifoliata

Tiny white flowers; forms carpets in redwood forests

ALUM ROOT
Heuchera micrantha

Tiny pink flowers, leaves tinted with red; in shady places

Fringe Cups
Tellima grandiflora

Red flowers, in moist places

ROSE FAMILY *Rosaceæ*

10 or more stamens; large family (3000 species) includes trees, shrubs and herbs

SILVERWEED
Potentilla egedii grandis

Yellow flower on low plant in wet places; undersides of leaves silvery

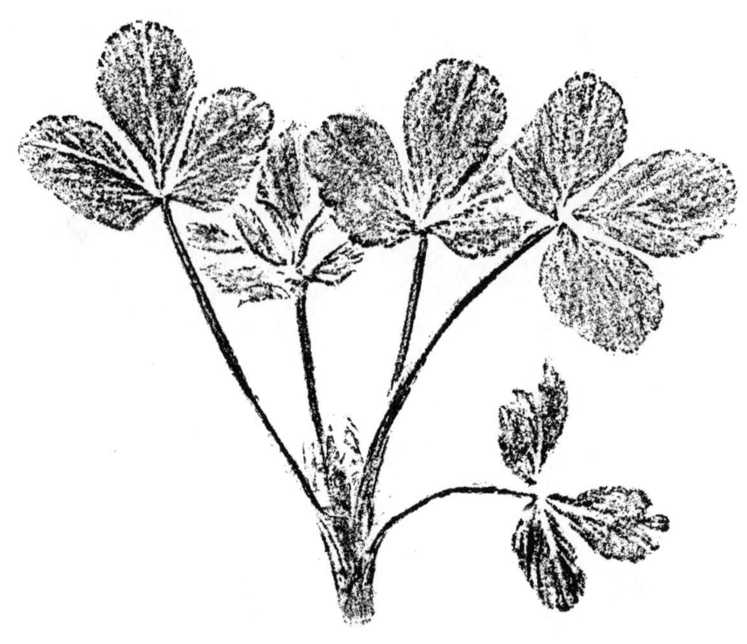

Beach Strawberry
Fragaria chiloensis

White flower, on dunes and bluffs

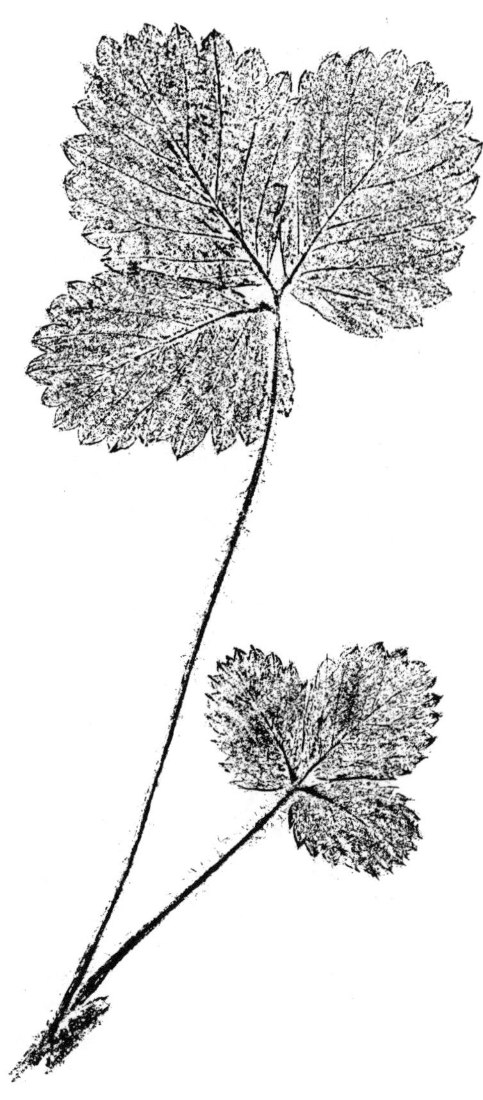

WOOD STRAWBERRY
Fragaria californica

White flower, on wooded hills

PEA FAMILY *Leguminosæ*

Large family (15,000 species) all bearing a common fruit, the 2-sided pea pod

False Lupine
Thermopsis macrophylla

Yellow flowers; only 3 palmate leaflets instead of 4 or more of true lupines. (Lupines all have palmately compound leaves and flowers in a raceme)

Sky Lupine
Lupinus nanus
Blue and white flower, grassy fields

Dove Lupine
Lupinus bicolor

Tiny blue and white flower, open areas

GULLY LUPINE
Lupinus densiflorus

White flower—or bluish or rose, extremely variable; twisted flower-stalk and round flat seed pods all on one side of stem.

CLOVER
Trifolium amplectens

Low plant, little red flower, all over

Sour Clover
Trifolium fucatum

Reddish or yellow flowers; all clovers have 3-leaflets

TOMCAT CLOVER
Trifolium tridentatum

Reddish flower, common all over

Indian Clover
Trifolium albopurpureum

Purple-red flower, abundant and widespread

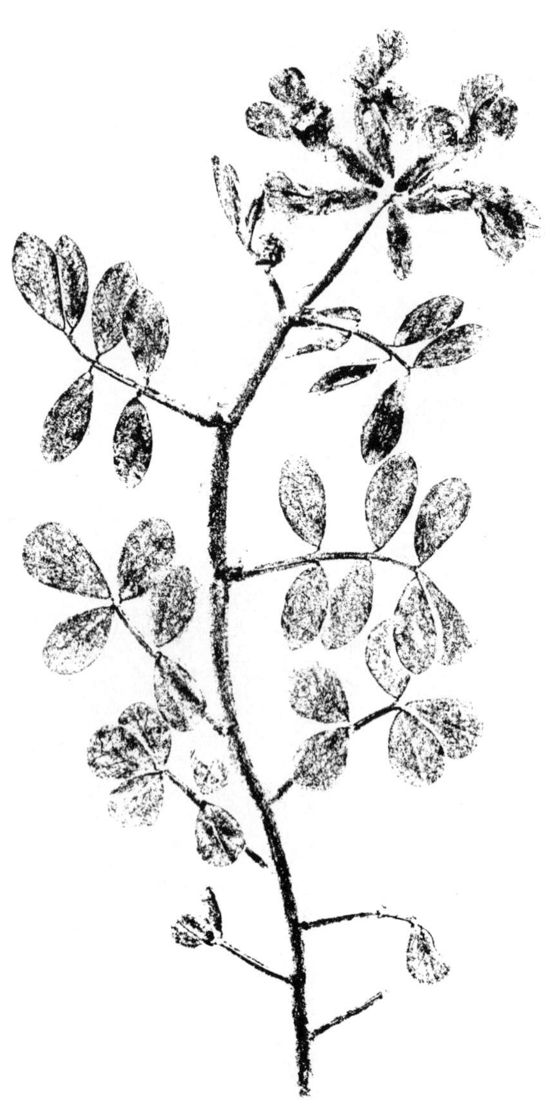

COAST TREFOIL
Lotus formosissimus

Yellow banner (upper petal), purple keel (lower petal)

Lotus corniculatus

Yellow flower; found all over (Introduced)

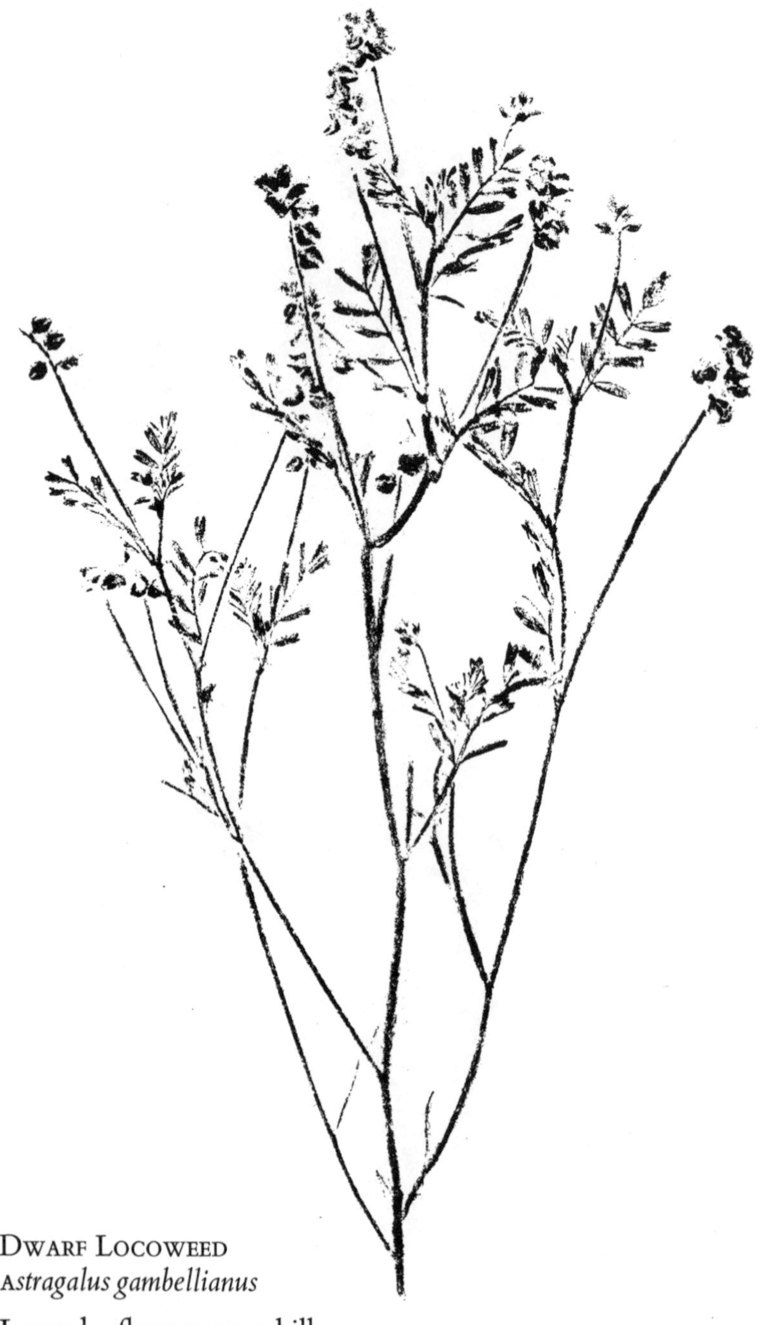

Dwarf Locoweed
Astragalus gambellianus

Lavender flower; open hills

SPRING VETCH
vicia sativa

Bright purple flower; vine-like herb (Introduced)

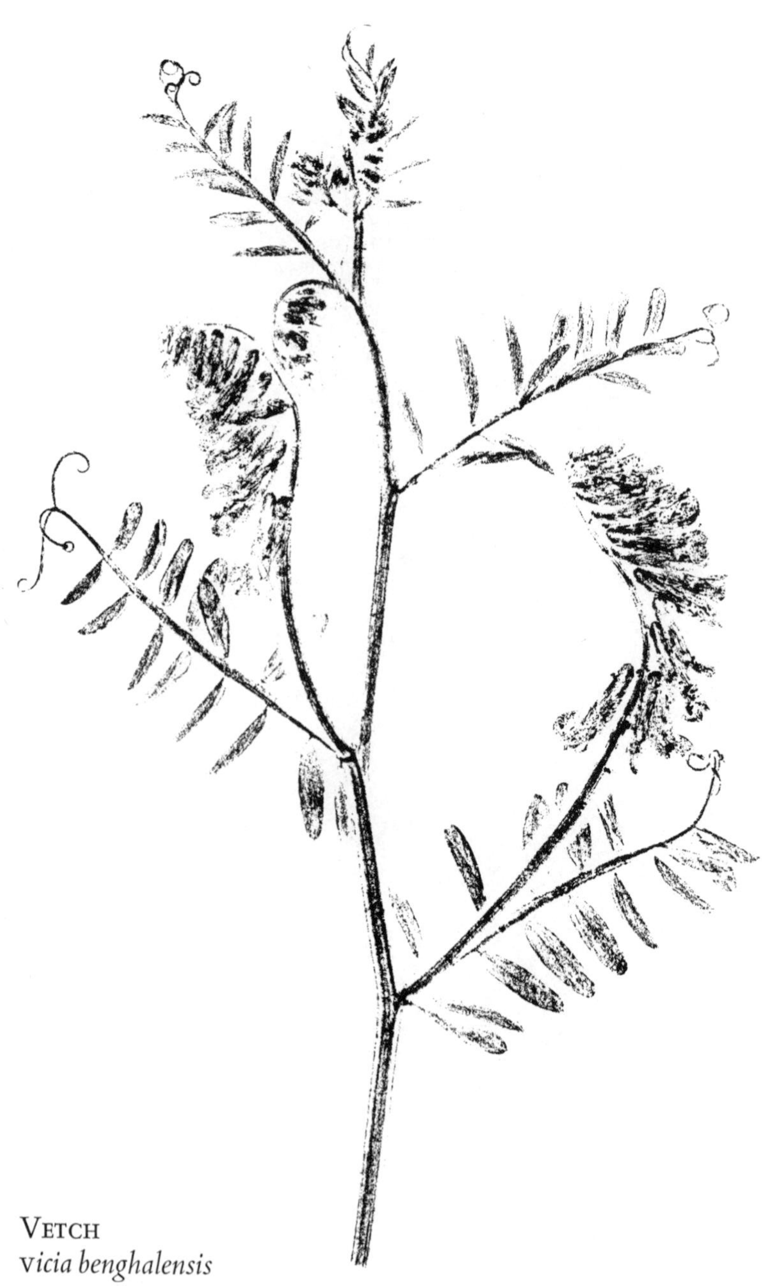

VETCH
vicia benghalensis
Bright purplish-red flower (Introduced)

WILD SWEET PEA
Lathyrus vestitus

Pink flower blooms in February

GERANIUM FAMILY *Geraniaceæ*

Flowers 5-petals, leaves sometimes deeply lobed; known as cranesbills, storkbills, heronsbills, clocks, filarees, because of long thin fruit.

REDSTEM FILAREE
Erodium cicutarium

Tiny purple blossom, long fruit (Introduced)

OXALIS FAMILY *oxalidaceæ*
Sour juice, leaflets in threes

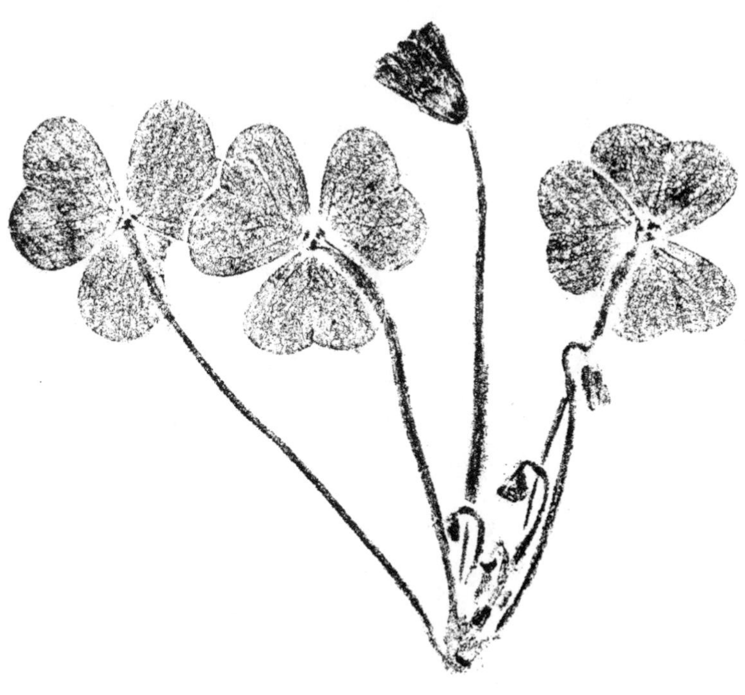

REDWOOD SORREL
oxalis oregano

Pink flower; in coast redwood forests

FLAX FAMILY *Linaceæ*

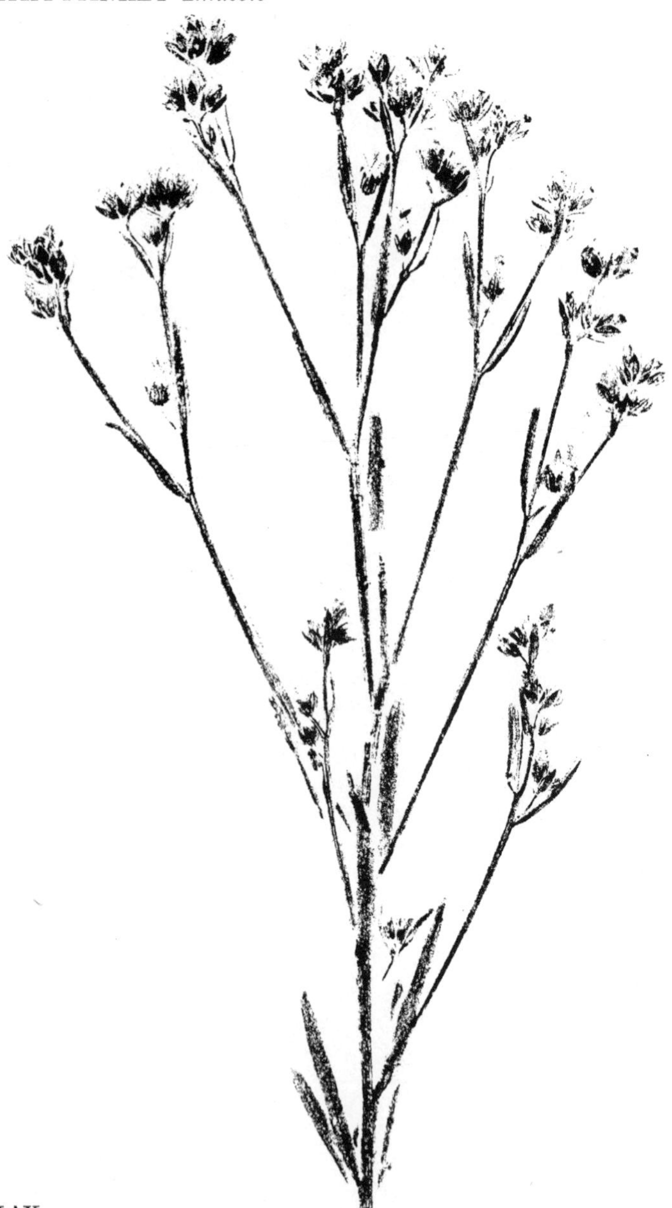

Flax
Linum congestum (*Hesperolinon congestum*)

Pink flower, San Francisco area only

MILKWORT FAMILY *Polygalaceæ*
Has a pea-like flower

Polygala californica

Purple flower, plant about a foot high

SPURGE FAMILY *Euphorbiaceæ*

TURKEY MULLEIN
Eremocarpus setigerus

Low gray-green leaves, dry places

MEADOW FOAM FAMILY *Limnanthaceæ*
5-petal flowers

Meadow Foam *Limnanthes douglasii*

5-petal spring flower, white with yellow center, often abundant on low wet ground

MALLOW FAMILY *Malvaceæ*

5-petal flowers have a tube-like center

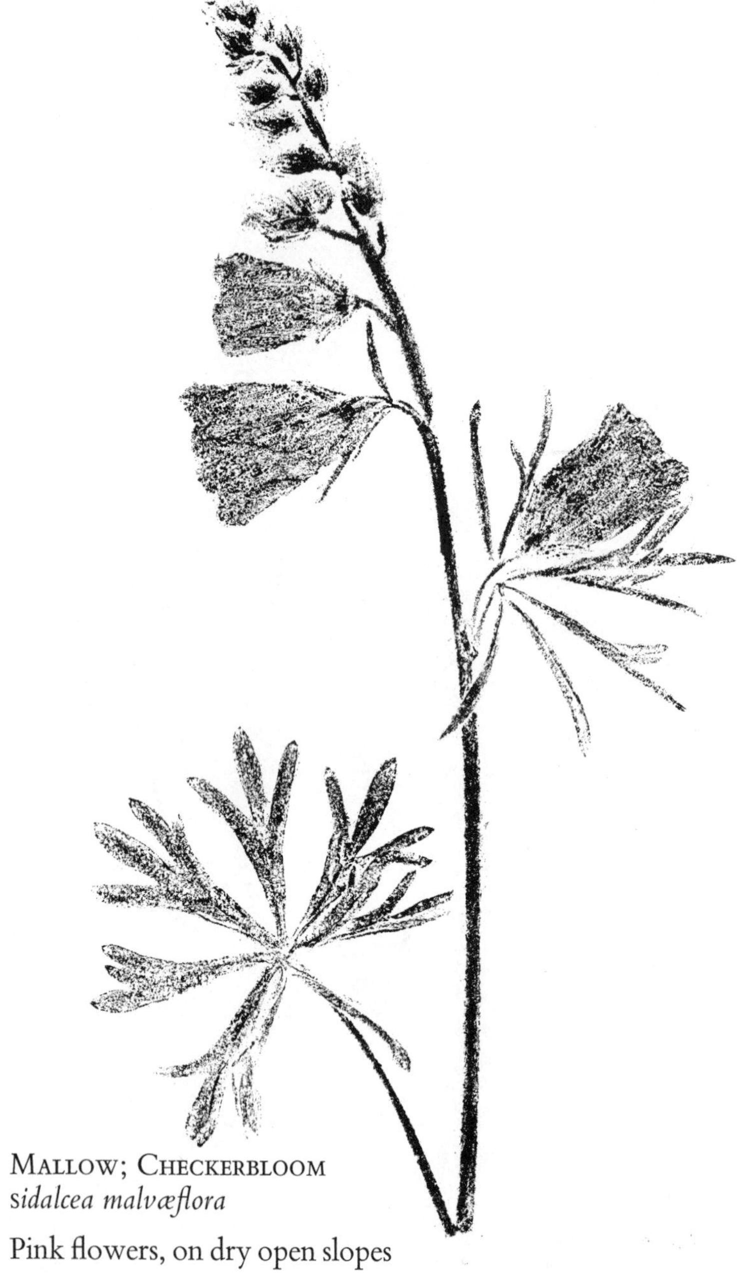

MALLOW; CHECKERBLOOM
Sidalcea malvæflora

Pink flowers, on dry open slopes

ST. JOHN'S WORT FAMILY *Hypericaceæ*

5-petals, opposite leaves

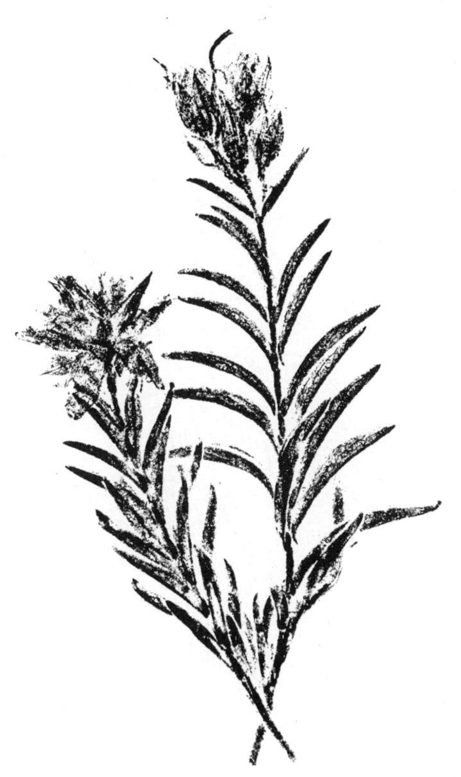

St. John's Wort
Hypericum concinnum

Yellow flower, in chaparral

FRANKENIA FAMILY *Frankeniaceæ*

FRANKENIA
Frankenia grandifolia
Tiny purple flower; salt marshes

VIOLET FAMILY *violaceæ*
Flowers with 5 petals, slightly irregular

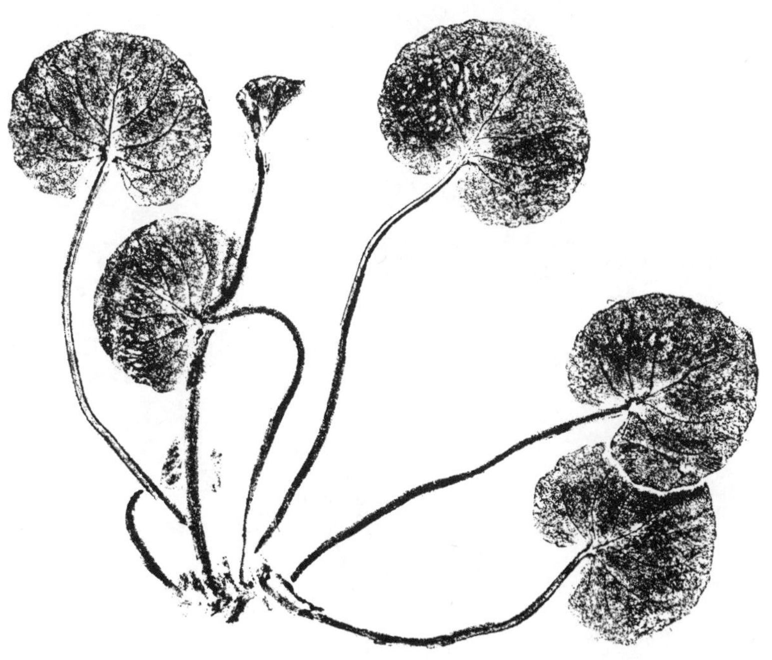

REDWOOD VIOLET
viola sempervirens

Yellow flower, near the coast and in coast redwood forests

JOHNNY-JUMP-UP
viola pedunculata

Yellow flower; on grassy hills

LOOSESTRIFE FAMILY *Lythraceæ*

LOOSESTRIFE
Lythrum hyssopifolia

Tiny purple flowers hidden at base of leaves; in wet places

EVENING PRIMROSE FAMILY *onagraceæ*

Flowers with 4 petals, usually regular

CALIFORNIA FUCHSIA
zauschneria californica

Bright red flowers; rocky places

FIREWEED; WILLOW-HERB
Epilobium franciscanum

Purple flower, long upright seed capsules; wet ground near coast

Lovely Clarkia
clarkia concinna

Cerise flower, in rocky banks

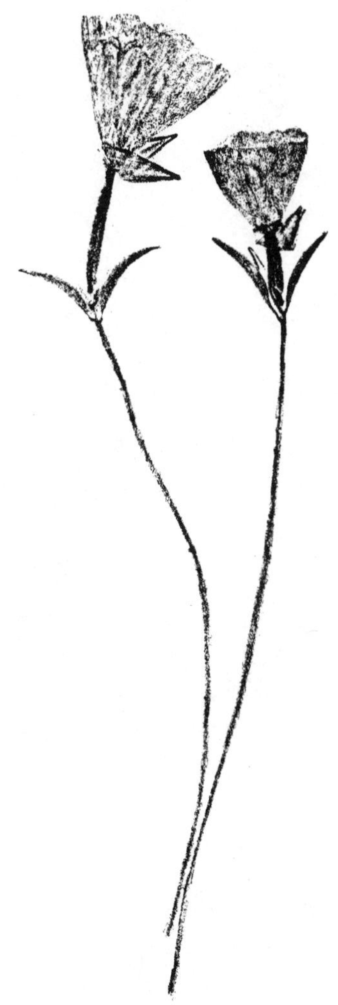

Farewell-to-Spring; Godetia
Clarkia rubicunda

Purple flower splotched inside with red; rocky places near coast

Evening Primrose
Oenothera hookeri

Large yellow flower opens at evening; near coast

CONTORTED SUN-CUP
Oenothera strigulosa

Sandy places and all over, near coast; yellow flower

SUN-CUPS
Oenothera ovata

Yellow flower, open places

ARALIA FAMILY *Araliaceæ*

Aralia; Elk Clover
Aralia californica

Plant grows to 10 feet in one season, from underground stems; always in shade, near streams; white flowers in umbels, red berries

PARSLEY FAMILY *Umbelliferæ*
Also called Carrot Family; large family with compound leaves, hollow ribbed stems, and flowers in clusters

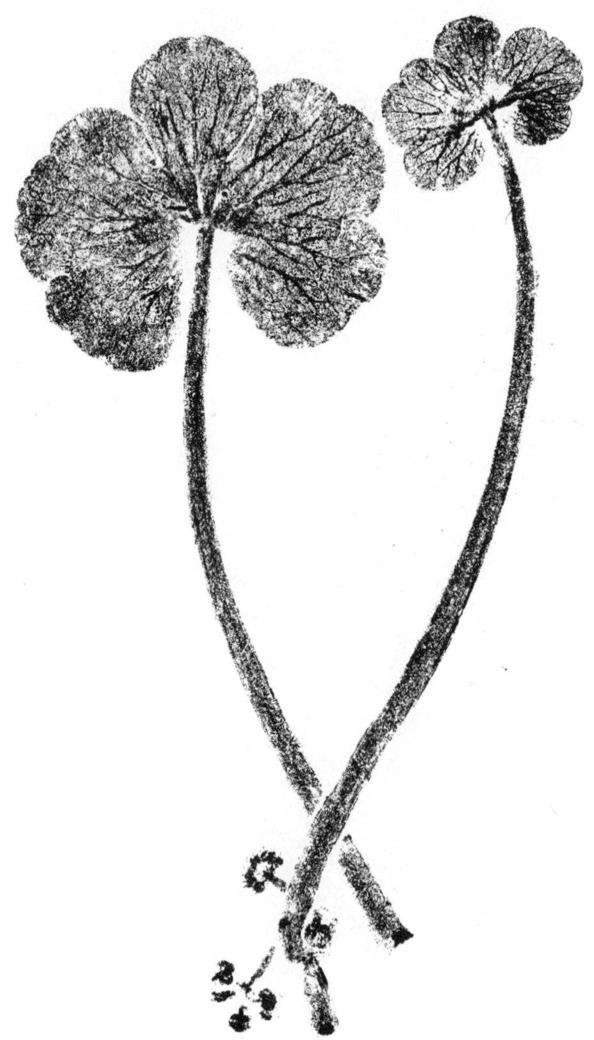

WATER PENNYWORT
Hydrocotyle ranunculoides

Clover-like leaves float on shallow ponds; tiny white flowers are at base of stems

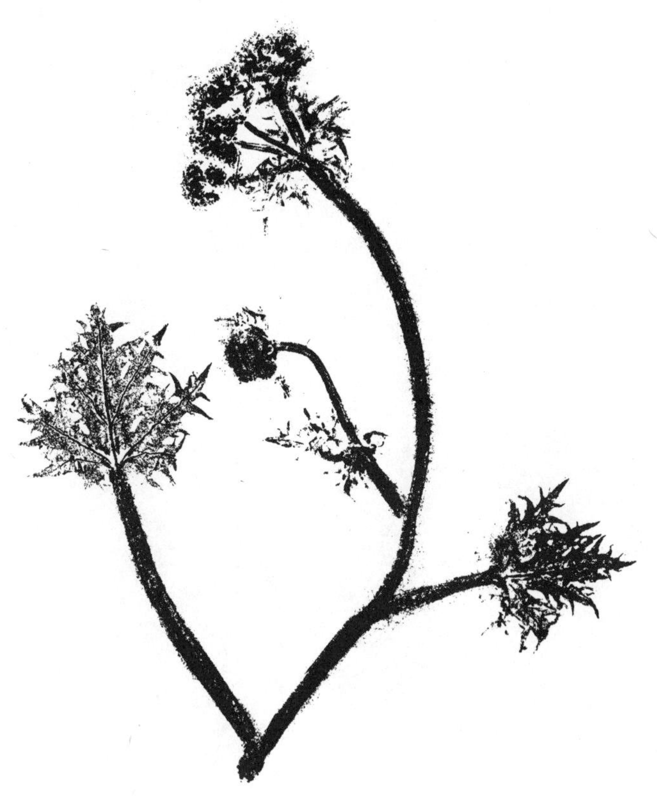

Coast Sanicle
sanicula laciniata

Yellow flowers, lacy leaves

Purple Sanicle
sanicula bipinnatifida

Common all over

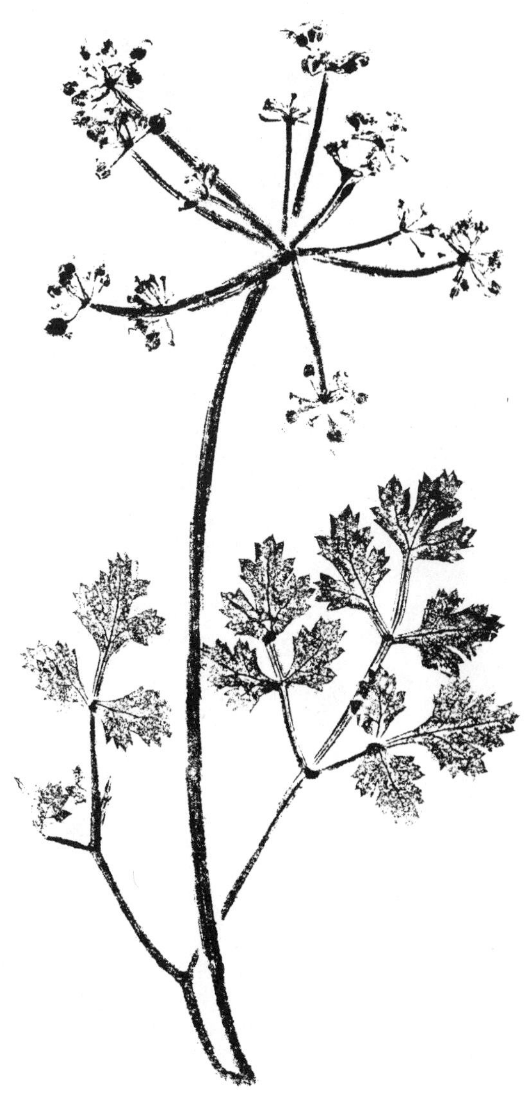

T*auschia kelloggii*

Yellow flowers, in shady places or in chaparral

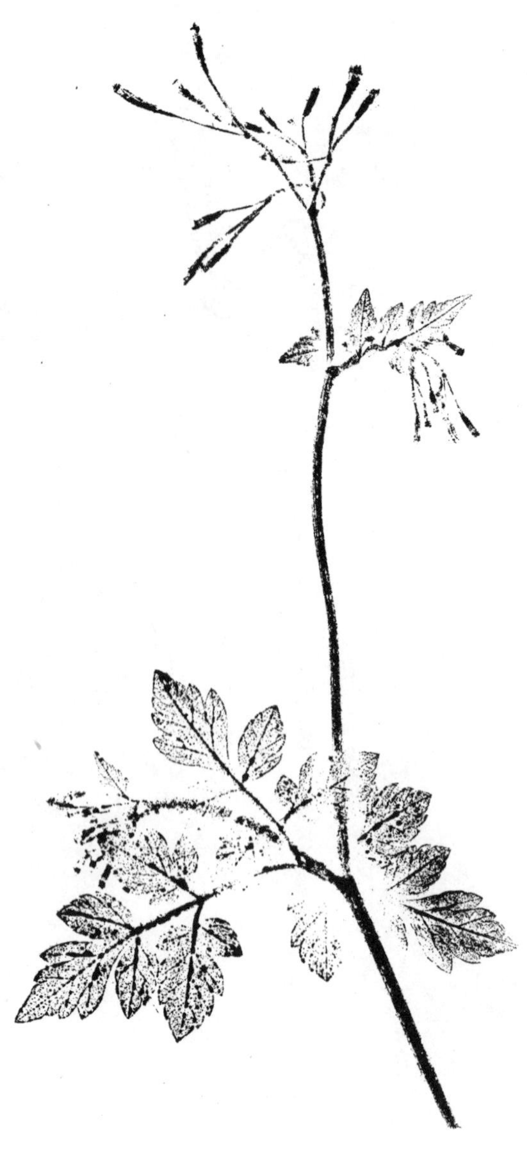

Sweet Cicely
Osmorhiza chilensis

Tiny white flower—pictured are seed pods; aromatic root

Queen Anne's Lace
Daucus carota

Pinkish flowers, along roadsides (Introduced)

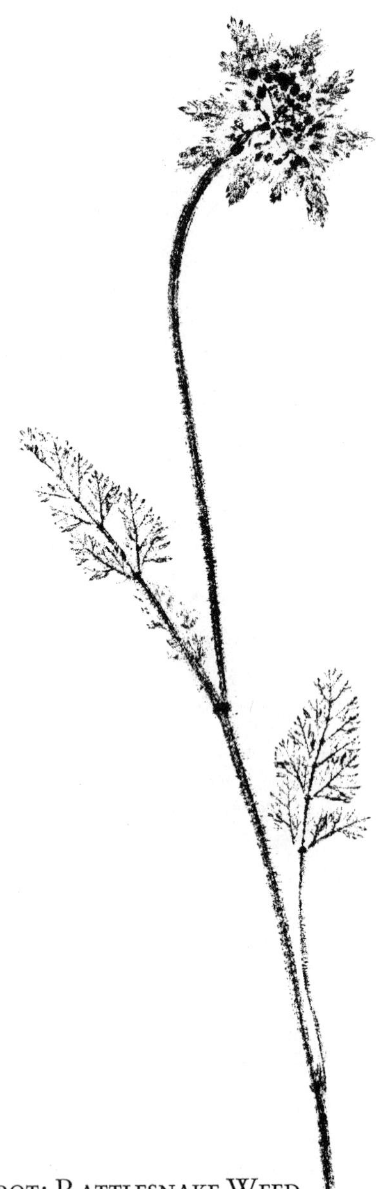

Wild Carrot; Rattlesnake Weed
Daucus pusillus

White flower, on grassy hills; once thought antidote for rattlesnake bite

SWEET FENNEL; ANISE
Foeniculum vulgare

Yellow flower, 6–8 feet tall along roadsides (Introduced)

POISON HEMLOCK
Conium maculatum

White flower, red-spotted stems, 6 feet, along roadsides; *poisonous* (Introduced)

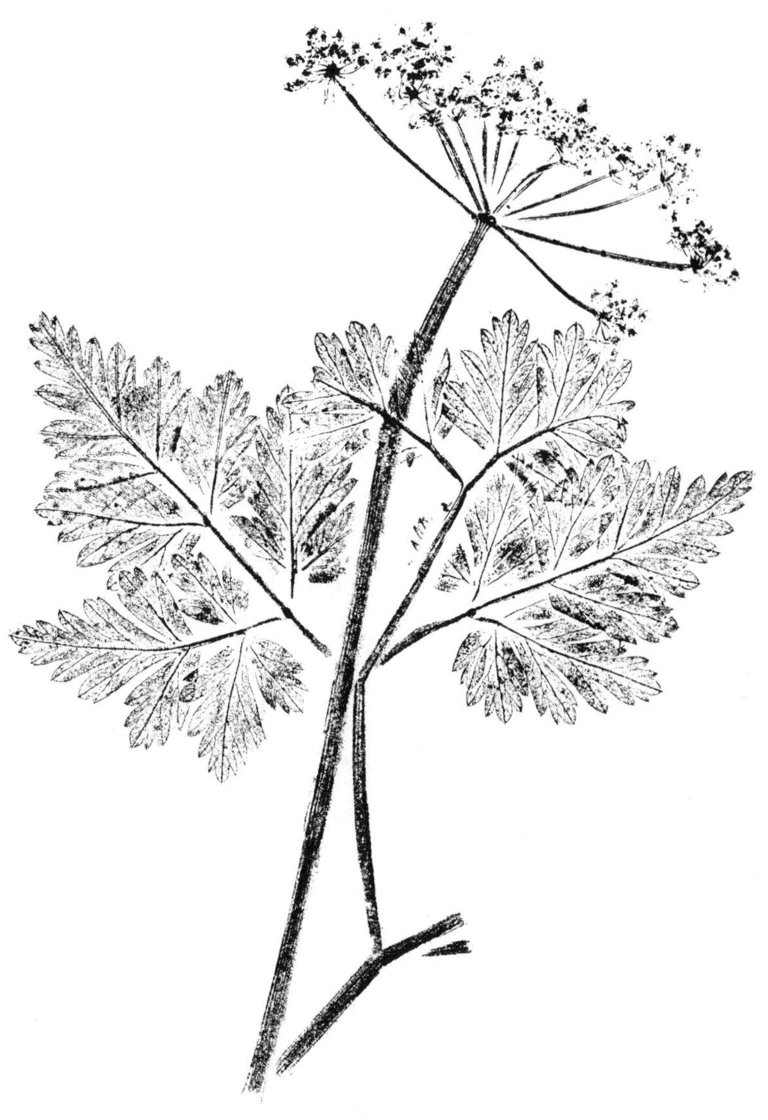

LOVAGE
Ligusticum apiifolium
White flower, in summer

Water Hemlock
Cicuta douglasii

White flower, 3 feet tall, near freshwater; *poisonous*

WILD PARSLEY
Oenanthe sarmentosa

White flower, damp places

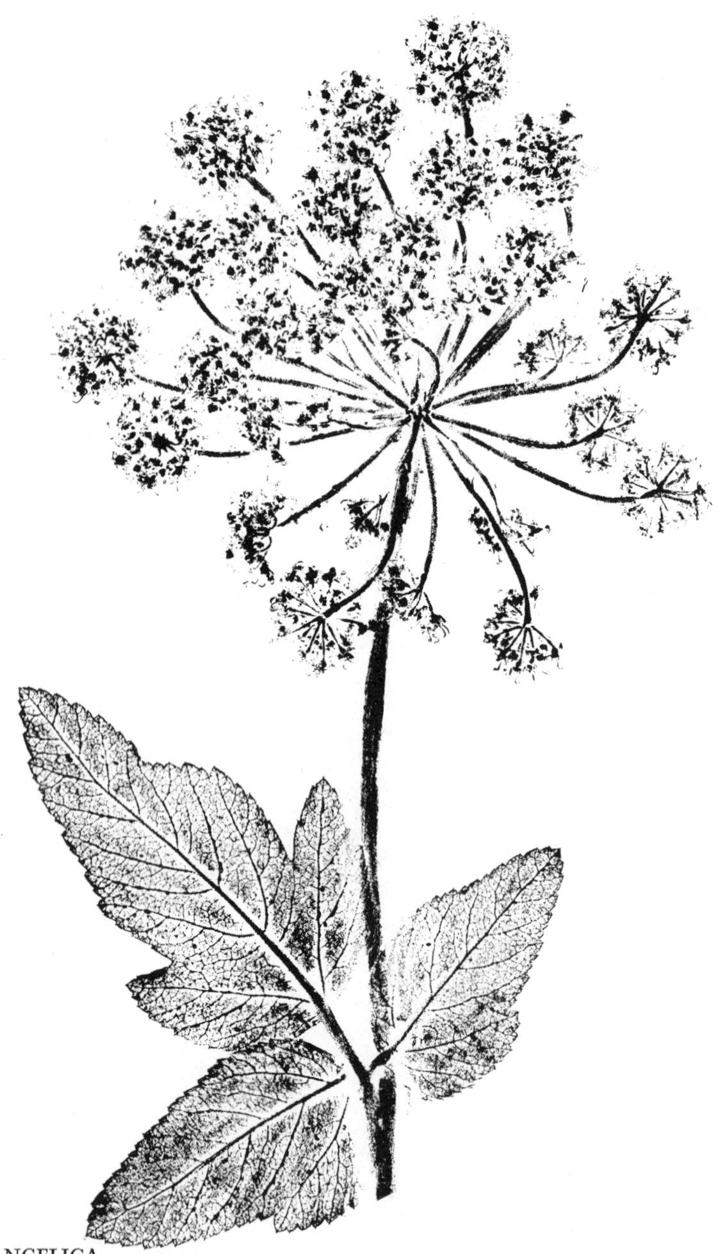

ANGELICA
Angelica tomentosa

White flower, in summer; large plant with compound leaves

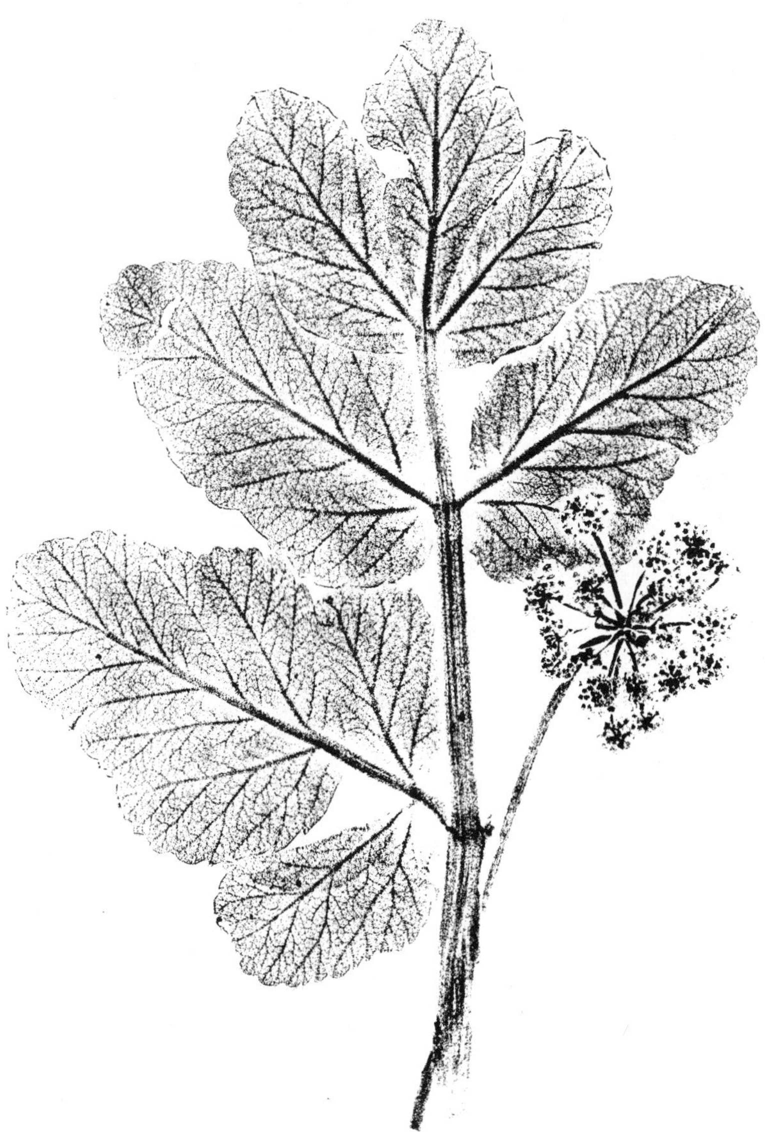

Coast Angelica
Angelica hendersonii

White flowers; undersides of leaves are white and wooly; near coast

Hog Fennel; Bladder Fennel
Lomatium sp.

Bright yellow flower, low plant

COW PARSNIP
Heracleum maximum

Cream flower on huge plant, almost always in bloom

COAST ERYNGO
Eryngium armatum

A spiny Umbellifera, in low places, near the ocean

PRIMROSE FAMILY *Primulaceæ*
Symmetrical flowers, with stamens opposite petals

STAR FLOWER
Trientalis latifolia

Tiny pink-purple flower, the size of a fly; shady places

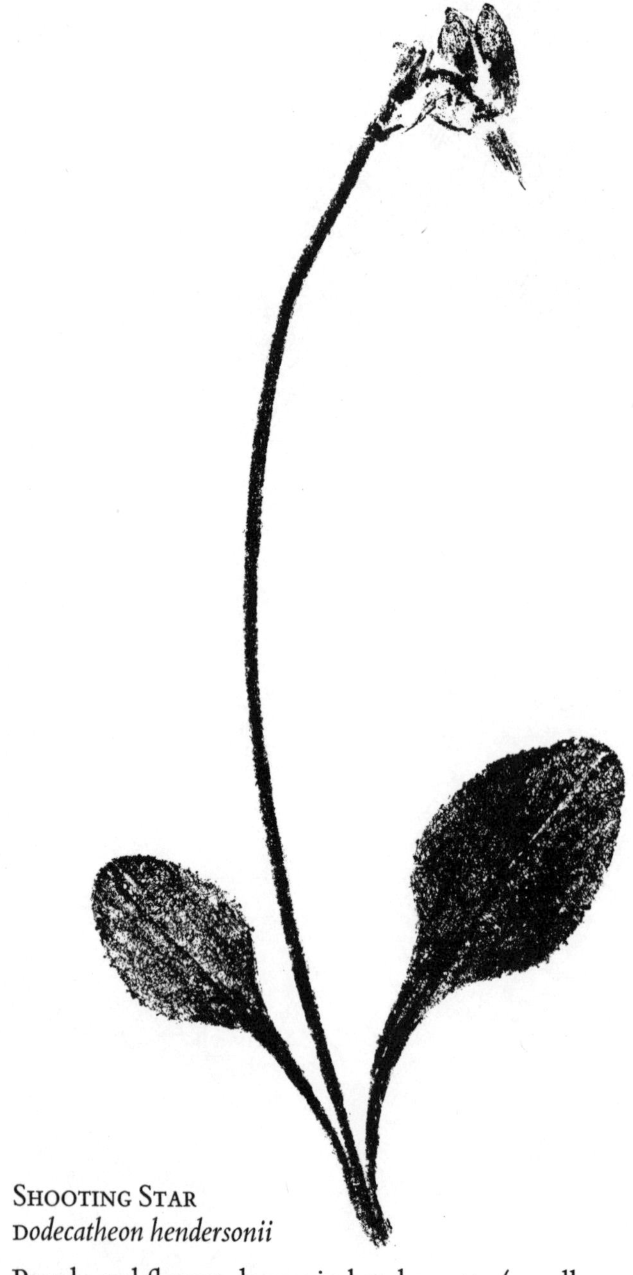

Shooting Star
Dodecatheon hendersonii

Purple-red flower, leaves in basal rosette (usually 2 or more flowers on flower stalk)

THRIFT FAMILY *Plumbaginaceæ*
(or Leadwort Family)

Sea Pink; Thrift
Armeria maritima

Pink flower clustered in tight head; on ocean bluffs

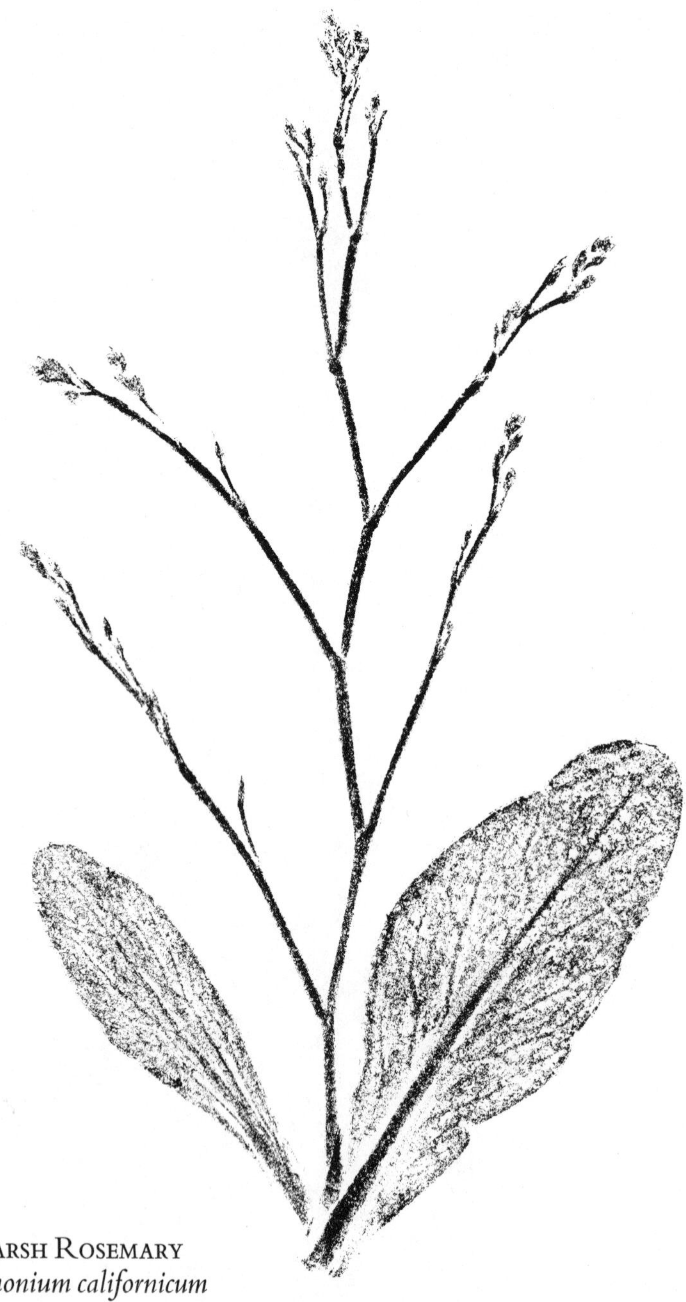

Marsh Rosemary
Limonium californicum

Tiny pink flowers; in salt marshes in summer

MORNING GLORY FAMILY *convolvulaceæ*

Trailing and twining plants

Morning Glory
convolvulus occidentalis

Red-and-white striped flowers on trailing stems

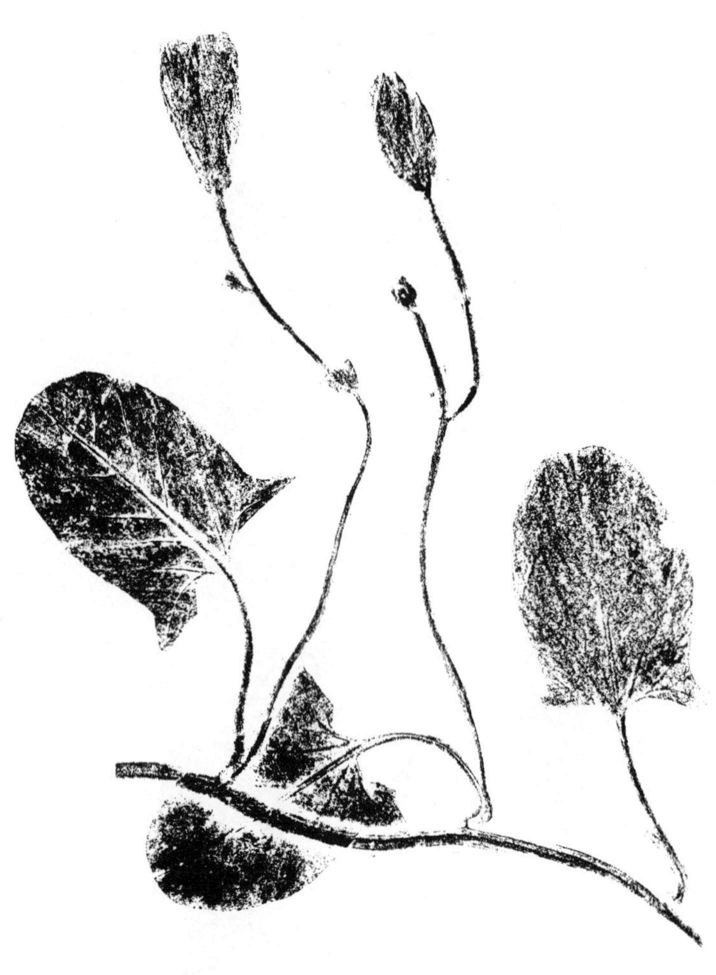

Bindweed
convolvulus arvensis

Pink flower, low-growing along roadsides (Introduced)

GILIA, OR PHLOX FAMILY *Polemoniaceæ*

Flowers mostly 5-lobed

LINANTHUS

Linanthus sp.

Blue flower, grassy places, common

NAVARRETIA
Navarretia rosulata

Blue flower, low-growing

WATERLEAF OR PHACELIA FAMILY *Hydrophyllaceæ*
Leaves usually hairy, divided flowers white to violet

BABY-BLUE-EYES
Nemophila menziesii

Blue flower, widespread in early spring

California Phacelia
Phacelia californica

Purple flower, blooms into fall, rocky places

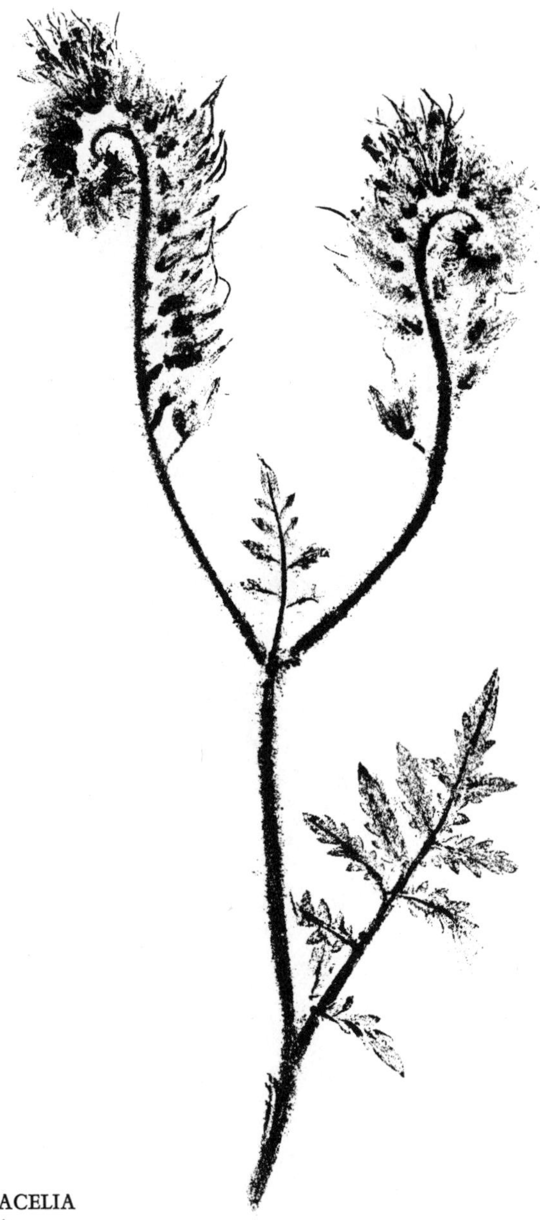

FERN PHACELIA
Phacelia distans

White flower, fern-like leaves

PHACELIA
Phacelia divaricata

Deep blue flower, in open places

BORAGE FAMILY *Boraginaceæ*

Small, rough and hairy plants, like phacelias, with flowers opening up from an unwinding coil

FIDDLENECK
Amsinckia spectabilis

Yellow flower, on sand dunes and near the coast

HOUND'S TONGUE
Cynoglossum grande

Purple-blue flower, pink sometimes, shady places early spring

MINT FAMILY *Labiatæ*
Square stems, opposite leaves, mint-like odor and irregular flowers, 2-lipped

SELF-HEAL
Prunella vulgaris

Purple flower, on coastal bluffs

Hedge Nettle
Stachys chamissonis

Purple flowers, large plant in wet places (not to be confused with stinging nettle, *Urtica sp.*)

HEDGE NETTLE NO. 2
stachys rigida

Rose-purple flowers; brushy, grassy, dry places

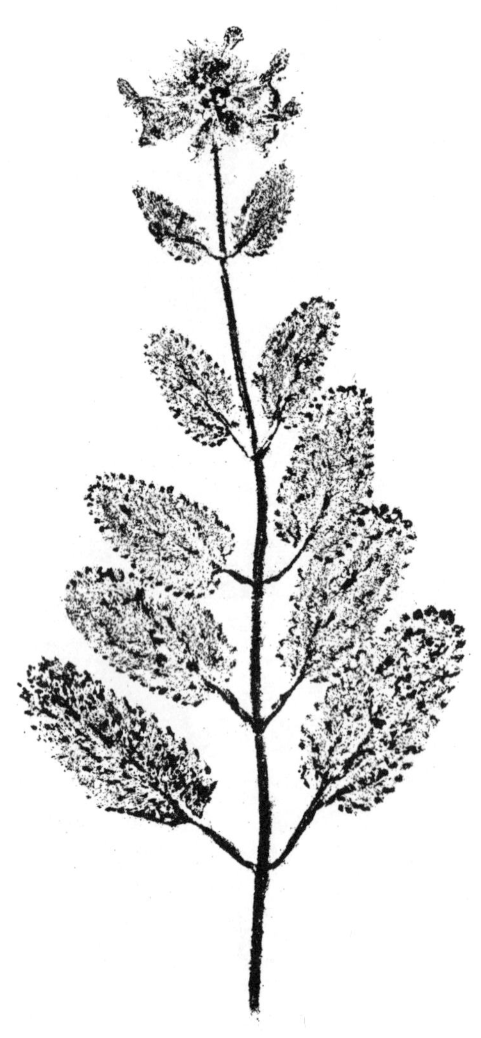

Hedge Nettle No. 3
Stachys pycnantha

Pink flower in round head, on serpentine

YERBA BUENA
satureja douglasii

White flower, fragrant, hugs the ground (makes fine tea)

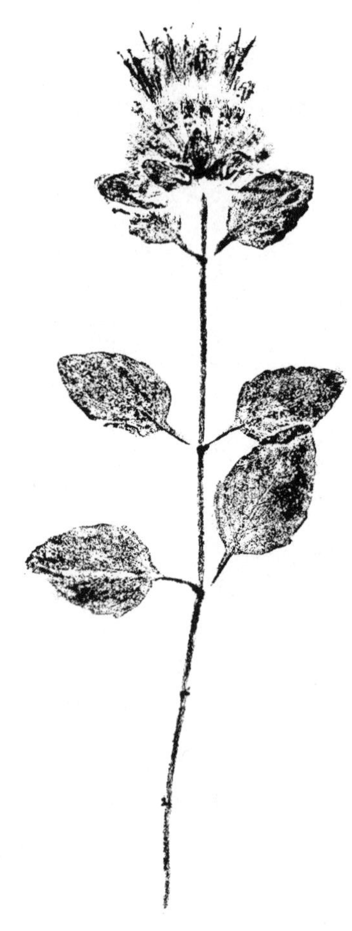

WESTERN PENNYROYAL
Monardella neglecta

Purple flower, on serpentine

NIGHTSHADE FAMILY *solanaceæ*

PURPLE NIGHTSHADE
solanum xanti var. *intermedium*

Shady or dry open places

Jimson Weed
Datura stramonium

Lavender flowers and prickly capsule; in summer; *poisonous*

FIGWORT FAMILY *scrophulariaceæ*
Large family with irregular flowers, many species in California

CHINESE HOUSES
collinsia heterophylla

Flower with white upper lip, purple lower lip, in whorls

California Bee-Plant
scrophularia californica

Small red flowers on tall stems, in many places

Mimulus
Mimulus guttatus

Yellow flower, wet places

DOUGLAS MIMULUS
Mimulus douglasii

Red-purple flower, chinless – no lower lip, plant 2 inches high, in gravelly soil

INDIAN WARRIOR
Pedicularis densiflora

Dark red flower, fern-like leaves

Indian Paintbrush
castilleja affinis

Orange flower, linear leaves, dry, rocky places; a partial parasite

Owl's Clover
Orthocarpus densiflorus

Pink flower, common on open hills

Purple Owl's Clover
Orthocarpus purpurascens

Showy and colorful purple stalks, in open places

Smooth Owl's Clover
Orthocarpus faucibarbatus

Yellow flower, in wet and swampy places, roadsides

BROOMRAPE FAMILY *orobanchaceæ*

Broomrape
orobanche fasciculata

Yellow flower; tiny plant, a root-parasite

PLANTAIN FAMILY *Plantaginaceæ*

Low plants with basal leaves, usually ribbed

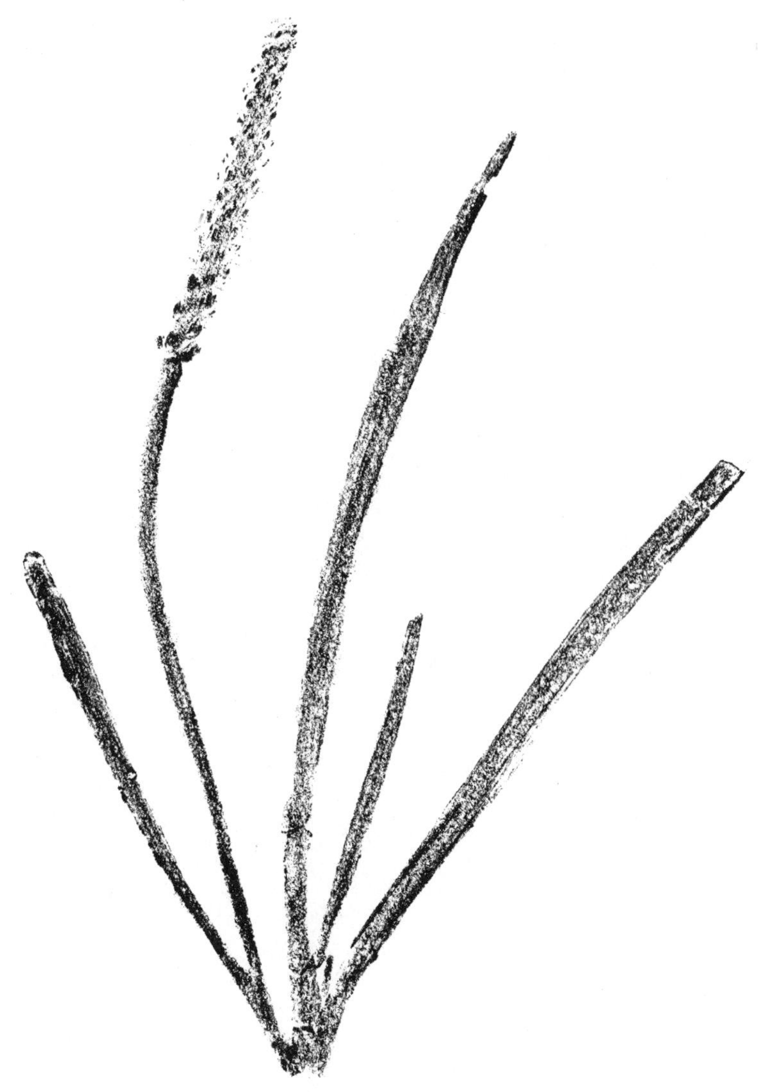

SEA PLANTAIN
Plantago maritima

Near ocean

Dwarf Plantain
plantago erecta

Tiny plantain, all over

MADDER FAMILY *Rubiaceæ*

BEDSTRAW
Galium sp.

Leaves in whorls, low-growing

VALERIAN FAMILY *Valerianaceæ*

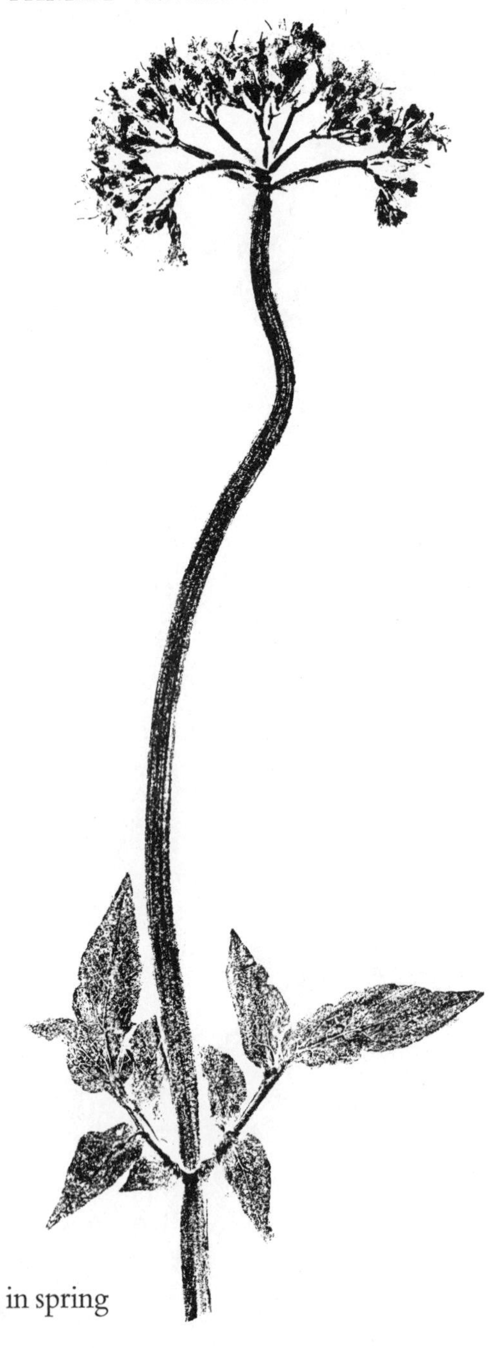

Plectritis sp.

Pink flower; in spring

CUCUMBER FAMILY *Cucurbitaceæ*

Family of gourds, pumpkins, melons, squashes, cucumbers

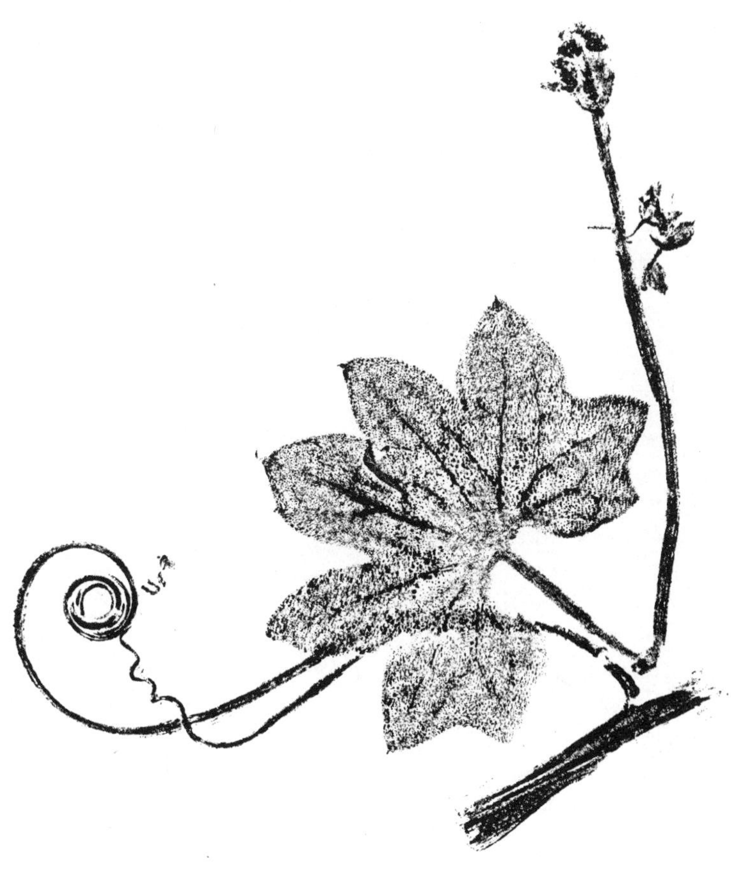

WILD CUCUMBER; MANROOT
Marah fabaceus

Vine-like plant with yellow flowers and large root

BELLFLOWER FAMILY *campanulaceæ*

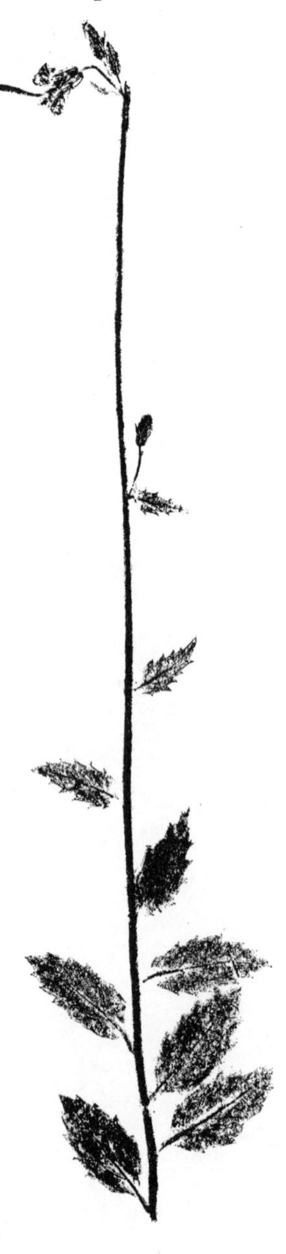

CALIFORNIA HARE-BELL
campanula prenanthoides

Small blue bell-like flowers on long slender stems; in woods

SUNFLOWER FAMILY *compositæ*

Two kinds of flowers, disk flowers in the center, ray flowers radiating out, hence name of Composite. Flowers are grouped into compact heads. A large family of 20,000 species

GRINDELIA; GUMWEED
Grindelia humilis

Yellow flower, near marshes

California Goldenrod
solidago californica

Yellow flowers on tall stems, late summer

COMMON ASTER
Aster chilensis

Pale blue flowers, late summer

Seaside Daisy
Erigeron glaucus

Purple rays, yellow disks, wooly leaves, only near the sea

Purple Cudweed
Gnaphalium purpureum

Purple flower, open places

Pearly Everlasting
Anaphalis margaritacea

Straw-colored flowers

Franseria
Franseria bipinnatifida

Yellow flowers in low clumps, sandy places, late summer and fall

MULE EARS
wyethia angustifolia

Large yellow flower, common, all over

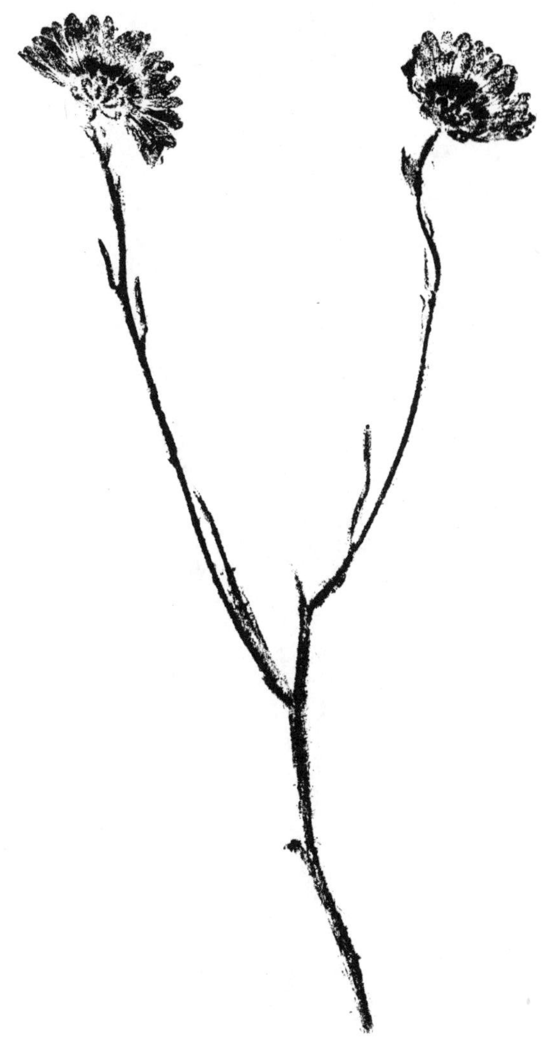

SPRING TARWEED
Hemizonia multicaulis

Yellow daisy-like flower with dark spots on disk flowers; blooms spring and fall

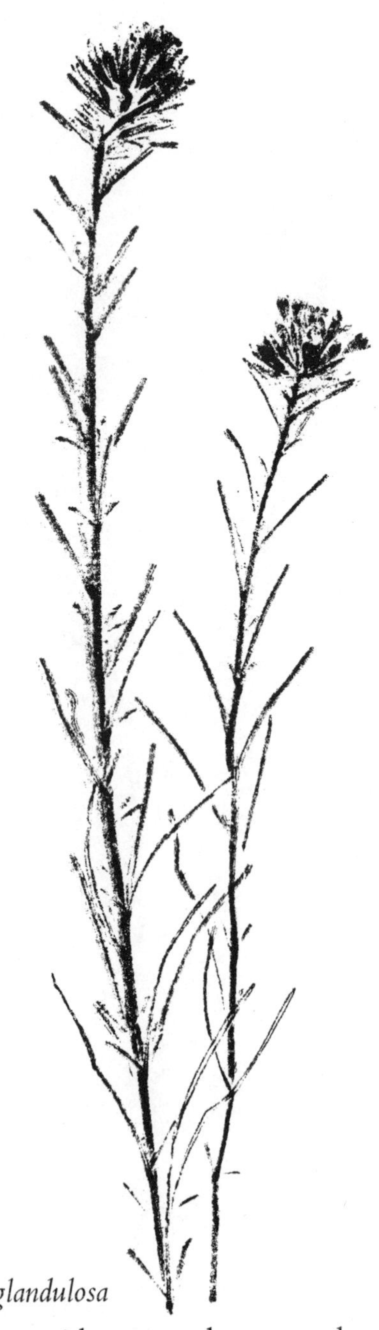

Rosin Weed
Calycadenia multiglandulosa
Pink flower, sticky, with strong odor; open places

Tidy Tips
Layia platyglossa

Yellow flower-heads, ray flowers tipped with white

Jaumea
Jaumea carnosa

Yellow flower in summer; in salt marshes; fleshy leaves

Gold Fields
Baeria chrysostoma

Orange-yellow flowers, in open places

Wooly Sunflower
Eriophyllum lanatum

Bright orange-yellow flower in summer; all over and highly variable

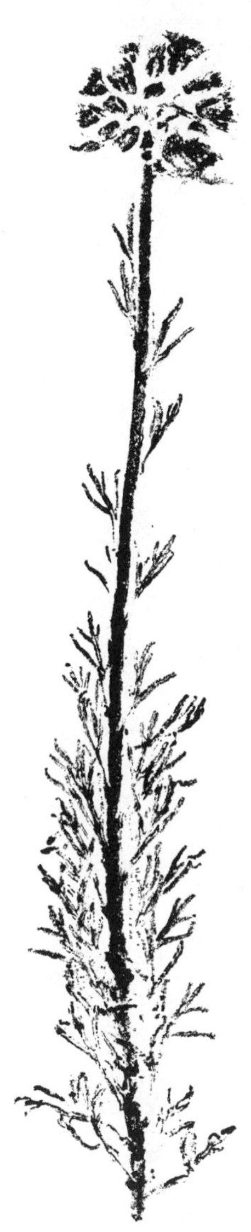

ERIOPHYLLUM
Eriophyllum confertiflorum

Golden flower, in dry places

Sneezeweed
Helenium puberulum

Brown and yellow flower, in moist places

YARROW
Achillea borealis

Whitish flowers in flat-topped clusters; blooms all year round

Brass Buttons
cotula coronopifolia

Bright yellow heads, wet places (Introduced)

California Mugwort
Artemisia douglasiana

Greenish flowers, aromatic, found all over

Sweet Coltsfoot
Petasites palmatus

White flower clusters on tall thick stems, large deeply lobed leaves; wet places

Coast Arnica
Arnica discoidea

Yellow heads; open places

Coastal Dandelion
Agoseris apargioides

Yellow heads, edible leaves

Hawkweed
Hieracium albiflorum

Delicate white flowers and hairy leaves

NOTES

Shrubs

SHRUBS are usually defined as low, woody plants with one or several stems growing from a clump. Trees are defined as woody plants with a tall trunk. But sometimes shrubs look like trees and often trees look like shrubs. If the plant you are trying to identify is not in one category, you may find it in the other. Chaparral areas are almost entirely shrubs; shrubs also grow in coast redwood and mixed forests and out on seacoast bluffs.

OAK FAMILY *Fagaceæ*

CALIFORNIA SCRUB OAK
Quercus dumosa

Shrub-like; small leaves, shiny and toothed

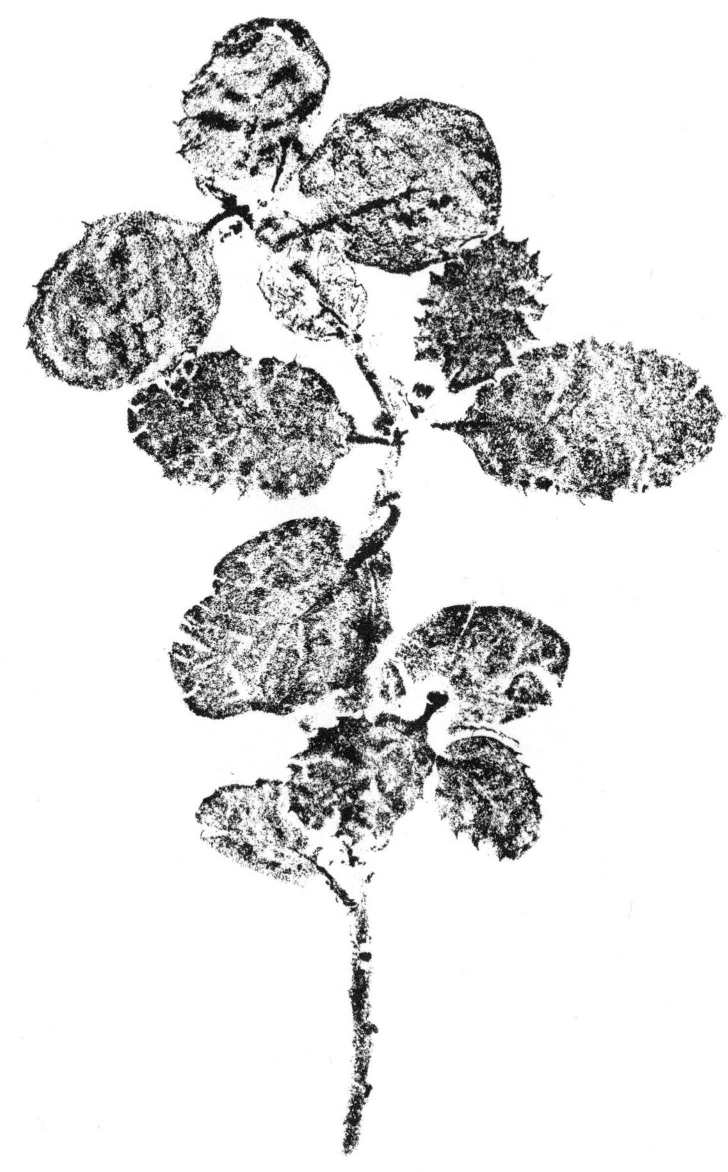

LEATHER OAK; SERPENTINE OAK
Quercus durata

Shrub-like; dull gray-green leaves, cupped; on serpentine only

BIRTHWORT FAMILY *Aristolochiaceæ*

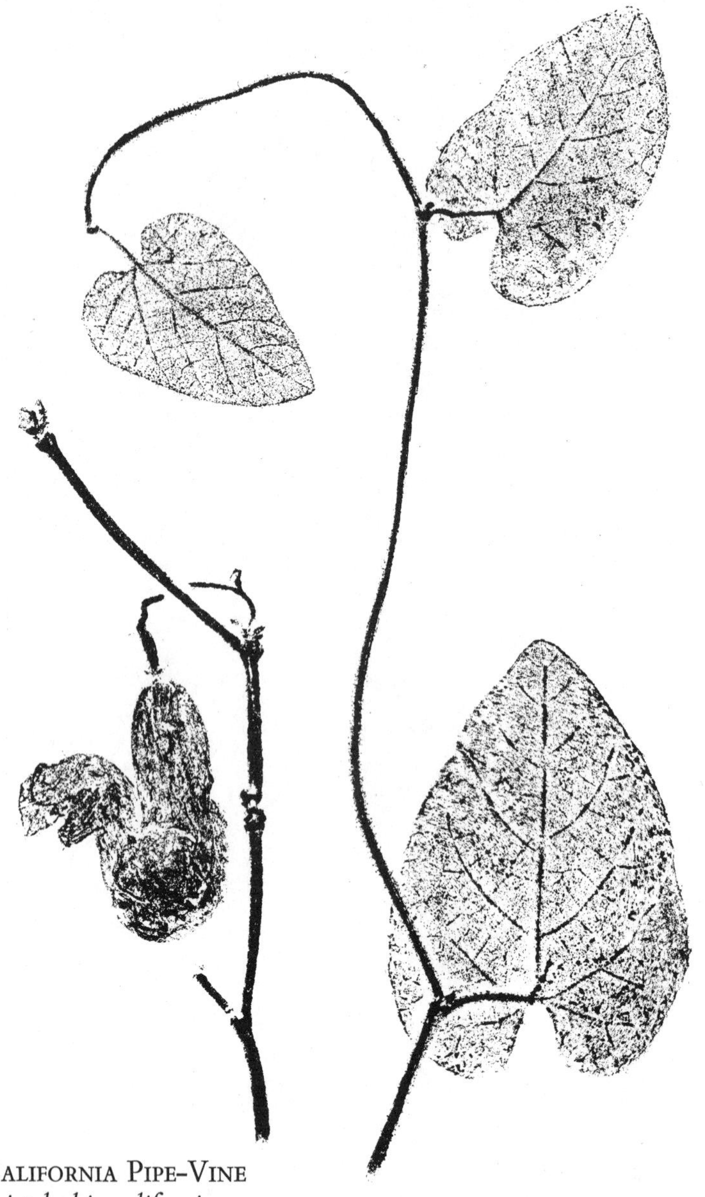

CALIFORNIA PIPE-VINE
Aristolochia californica

Pipe-shaped flowers appear in January, leaves come out later.

BUTTERCUP FAMILY *Ranunculaceæ*

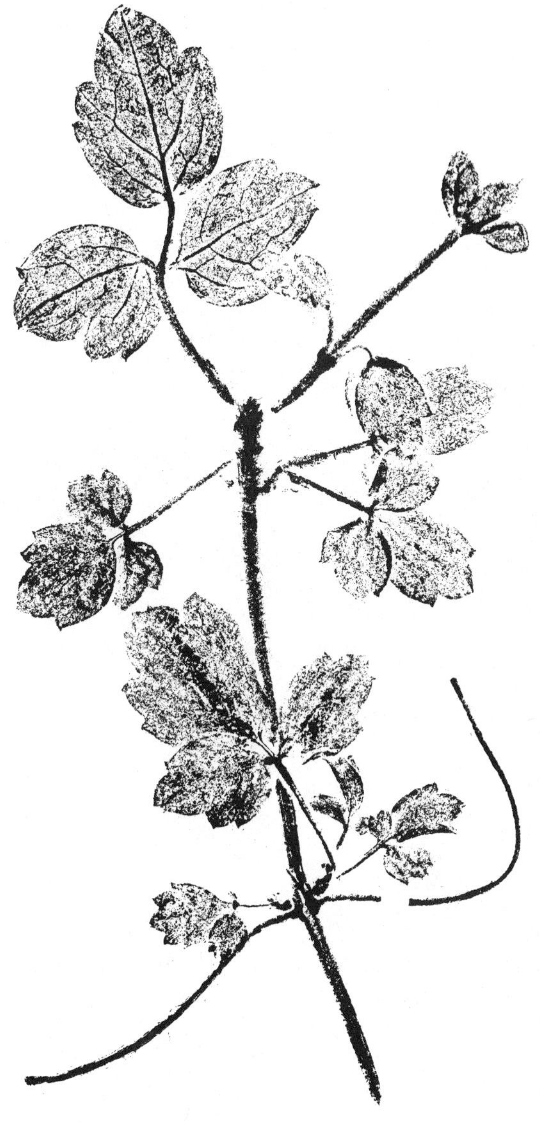

Clematis lasiantha

Creamy flower; in chaparral

SWEET SHRUB FAMILY *Calycanthaceæ*

SPICE BUSH
Calycanthus occidentalis

Red chrysanthemum-like flower with spicy odor; near water

BARBERRY FAMILY *Berberidaceæ*

MAHONIA; OREGON GRAPE
Mahonia pinnata

Yellow flower and blue berries

POPPY FAMILY *Papaveraceæ*

Tree Poppy
Dendromecon rigida

Yellow flower; in chaparral

SAXIFRAGE FAMILY *saxifragaceæ*

MODESTY
whipplea modesta

White flower; a trailing shrub in redwood forests

Flowering Currant
Ribes glutinosum

Pink flower

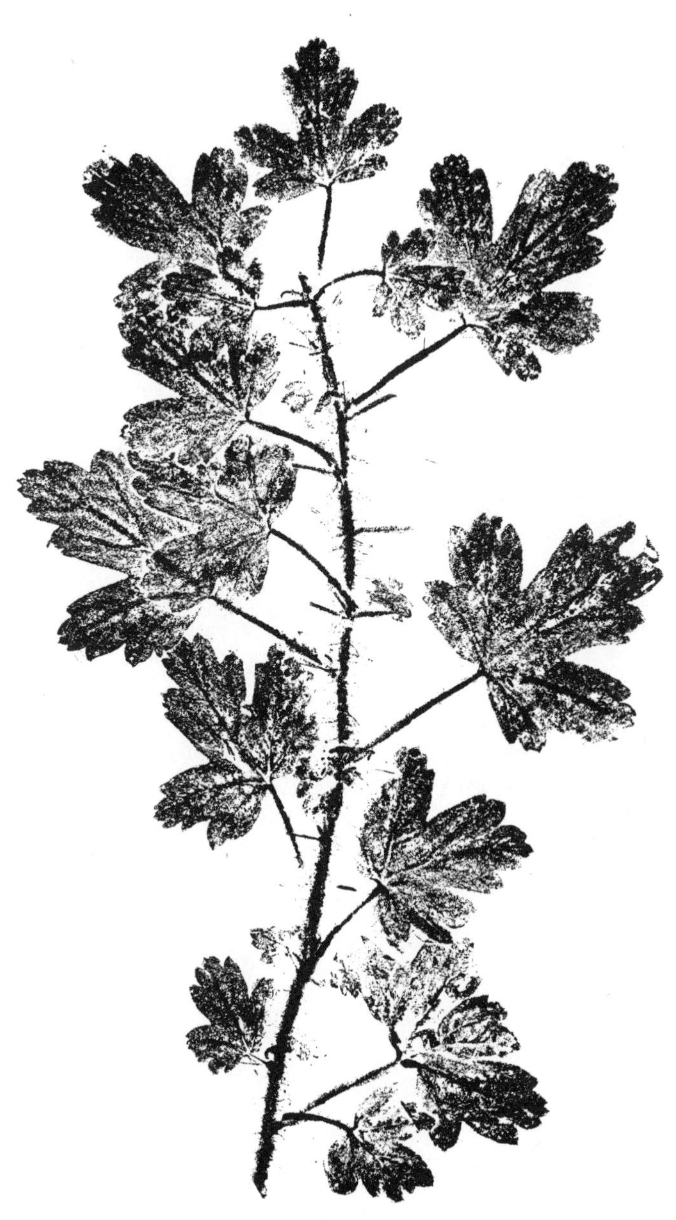

Coast Gooseberry
Ribes menziesii

Purple flower

ROSE FAMILY *Rosaceæ*

10 or more stamens; large and diverse family includes trees, shrubs and herbs. (3000 species)

WESTERN NINEBARK
Physocarpus capitatus

White flower, deciduous; creeksides and damp places

OCEAN SPRAY; CREAM BUSH
Holodiscus discolor

Cream colored flower

TOYON; CHRISTMAS BERRY
Photinia arbutifolia

White flower and red berries; evergreen

SERVICE BERRY
Amelanchier pallida

White flower, deciduous; rocky outcrops at edge of woods

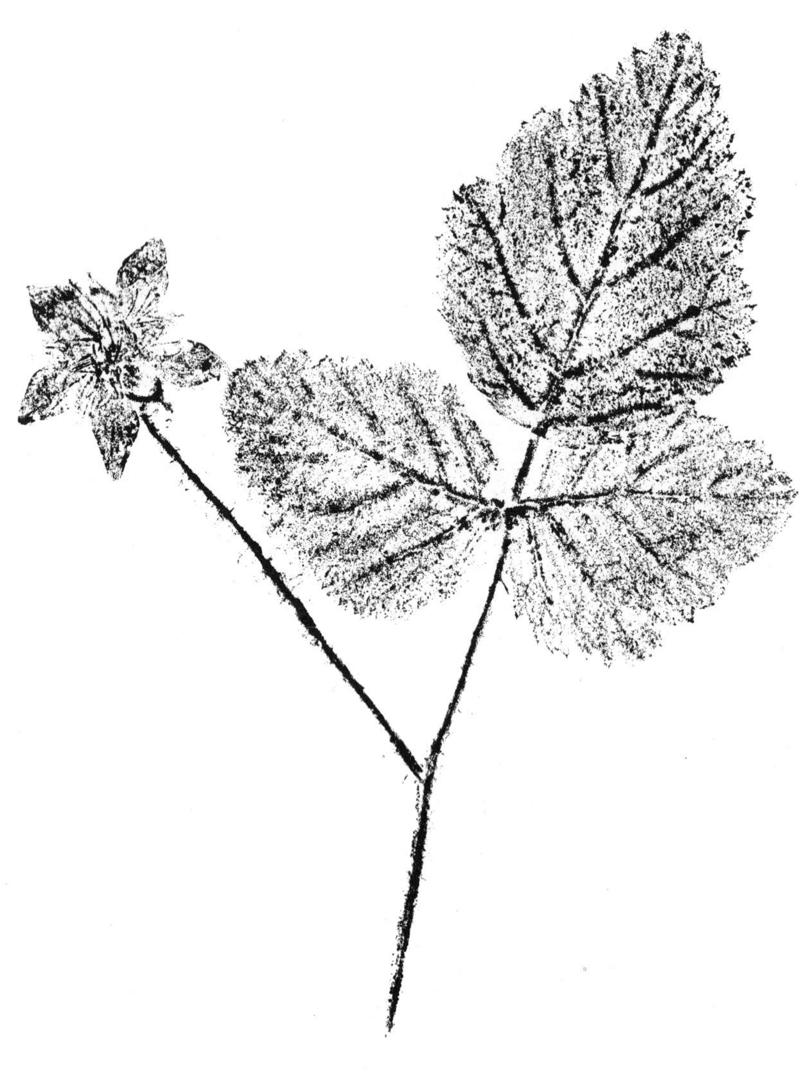

California Blackberry
Rubus ursinus

White flower, edible fruits; 3 leaflets, evergreen; thorns

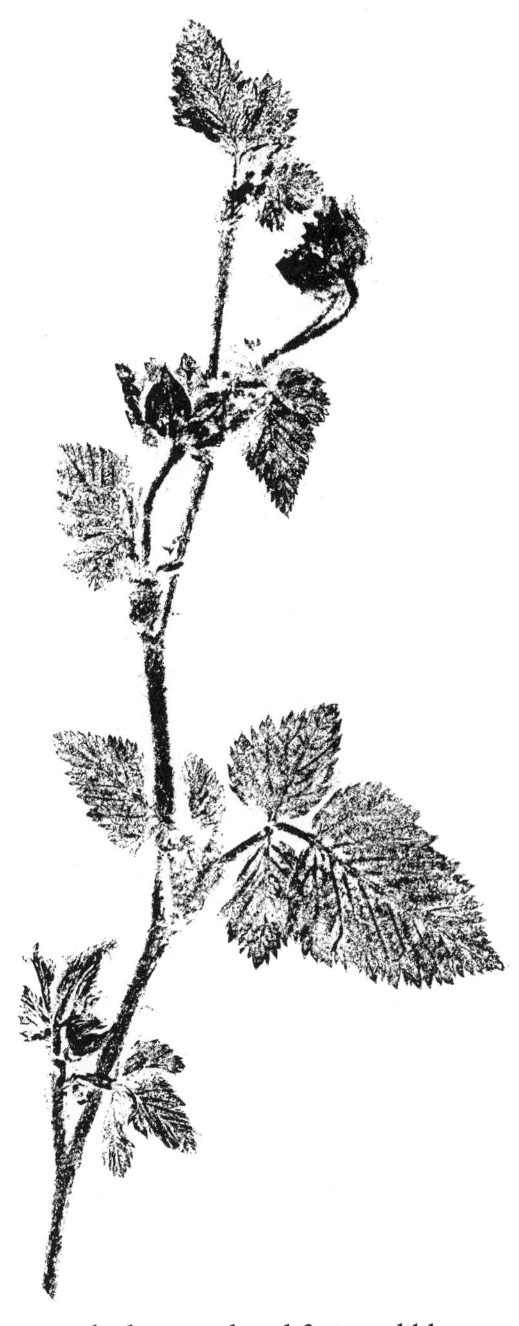

SALMONBERRY
Rubus spectabilis

Red-purple blossoms and salmon-colored fruits; edible

THIMBLEBERRY
Rubus parviflorus

White flower and thimble-shaped red berries; edible

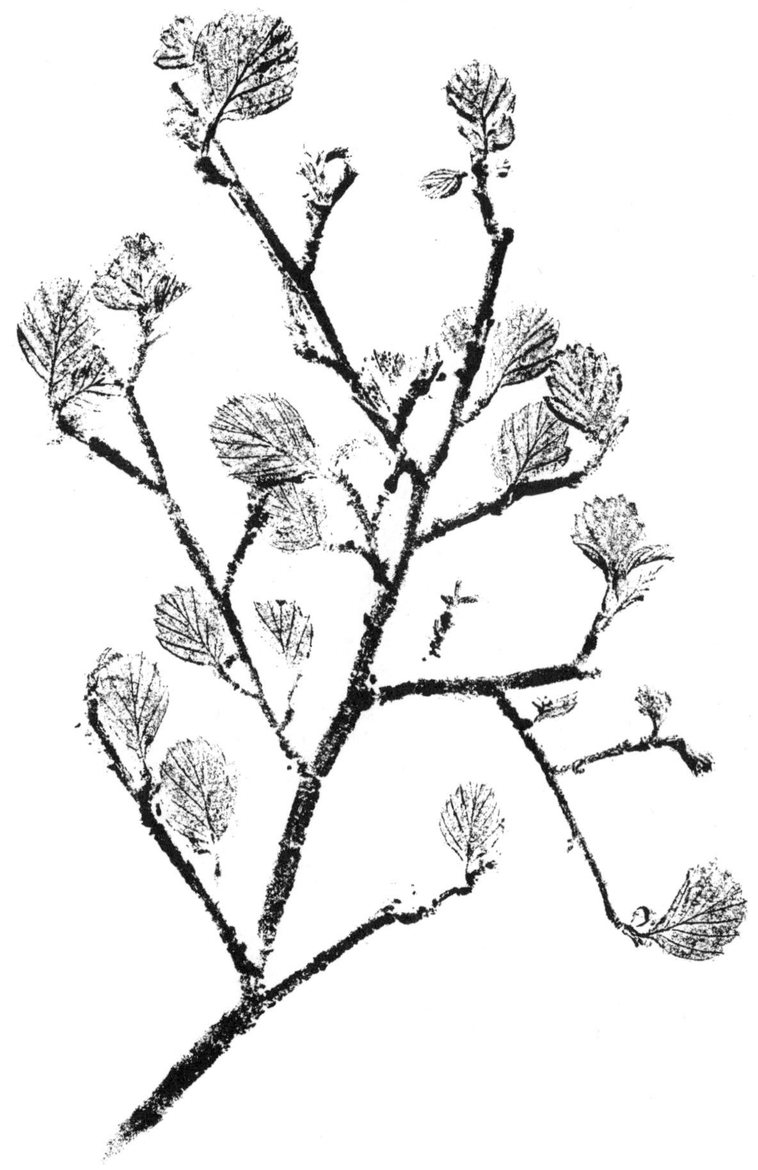

MOUNTAIN MAHOGANY
Cercocarpus betuloides

Brushy and dry hillsides, inner and outer Coast Ranges

Chamise; Greasewood
Adenostoma fasciculatum

White flower; only in chaparral

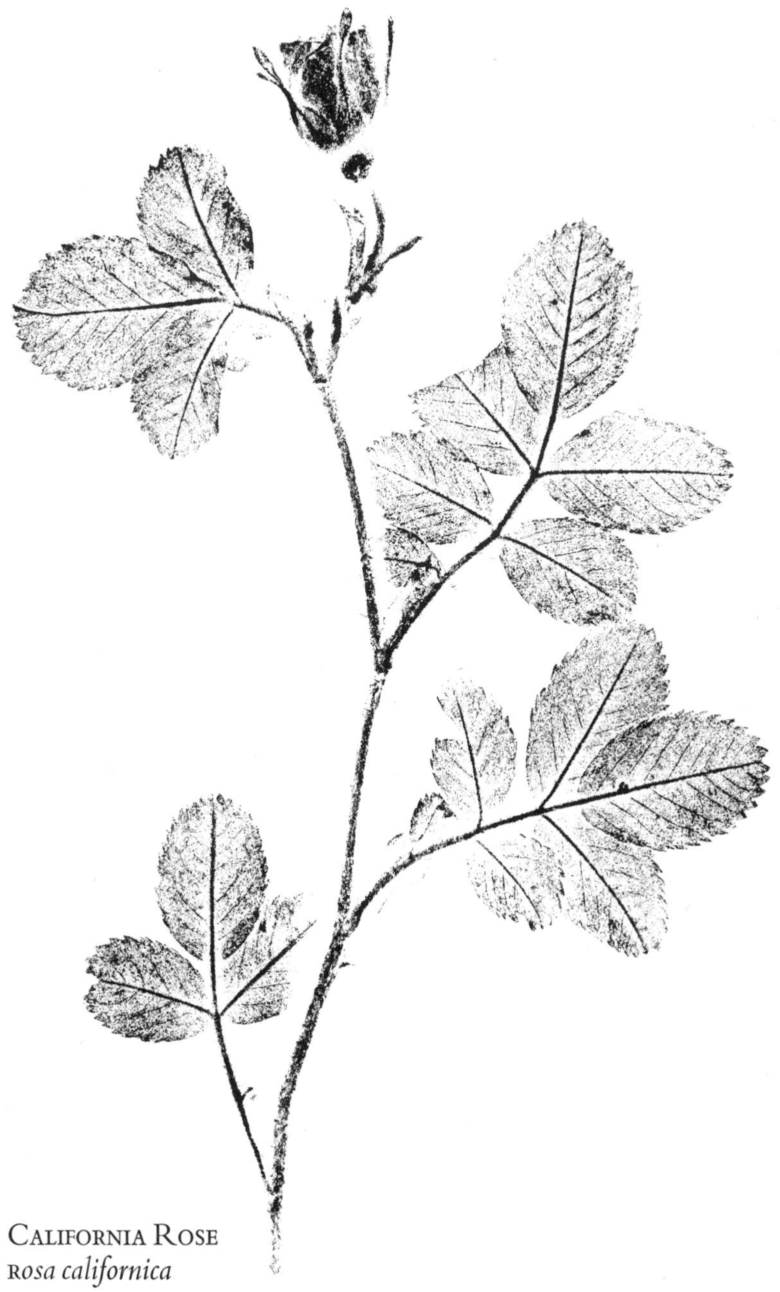

California Rose
rosa californica

Pink-lavender flowers (bud shown here); near water

Wood Rose
Rosa gymnocarpa

5-petal flower a deep bright pink

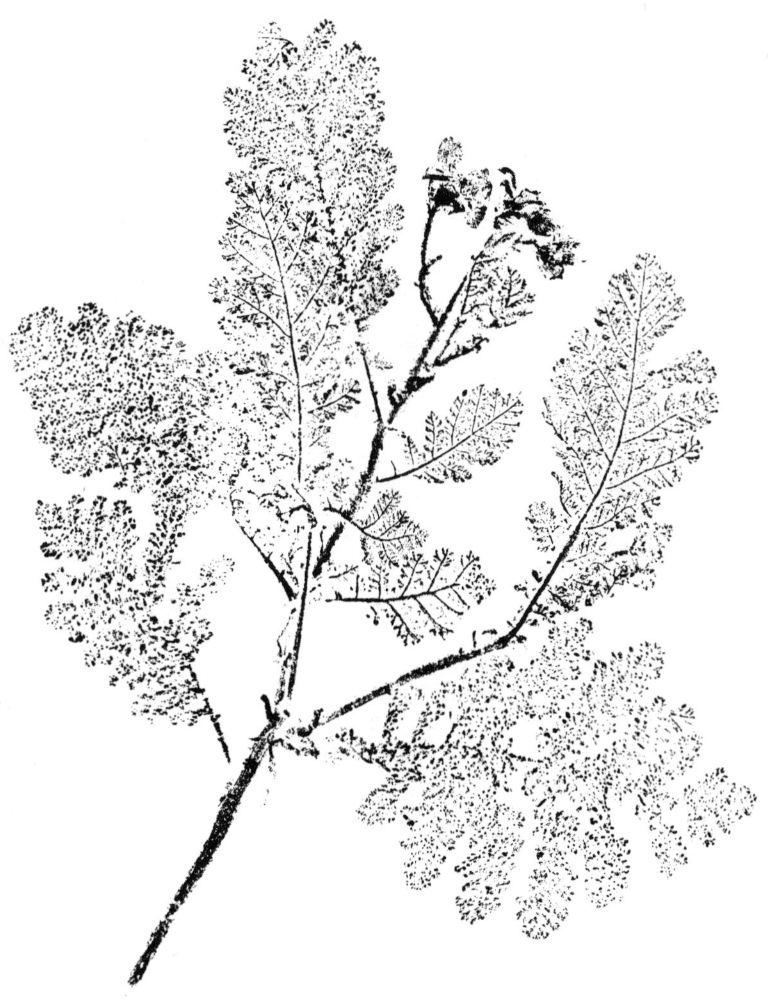

MOUNTAIN MISERY; BEAR CLOVER; KIT-KIT-DIZZE
Chamæbatia foliolosa

White flower and fern-like leaves; Sierra Nevada

Oso Berry
Osmaronia cerasiformis
White flower, deciduous

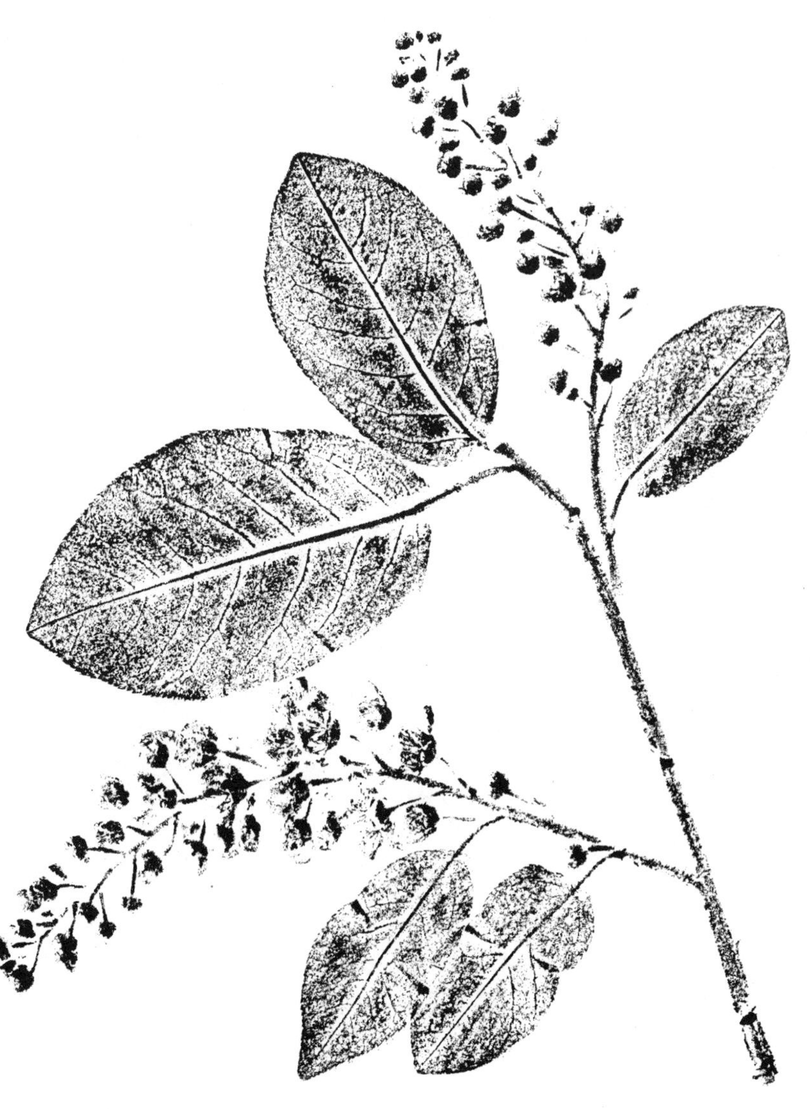

WESTERN CHOKECHERRY
prunus demissa

White flower, deciduous; creeksides

PEA FAMILY *Leguminosæ*

Large family, 15,000 species, divided into three subfamilies, all bearing a common fruit, the 2-sided pea-pod or a modification of it

CHAPARRAL PEA
Pickeringia montana

Red-purple flower and thorns

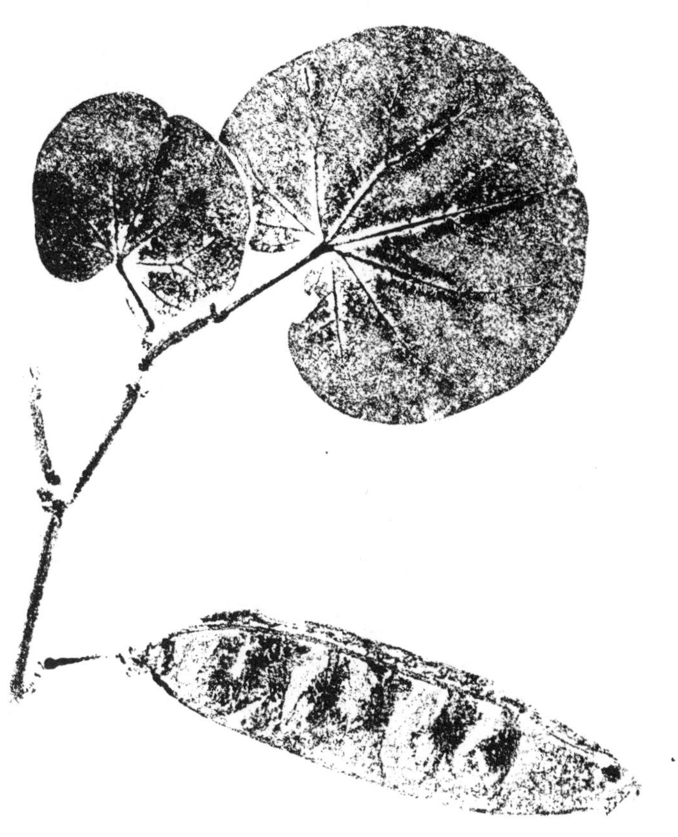

WESTERN REDBUD
Cercis occidentalis

In leaf, in July, and in fruit — a flat pod

Yellow Bush Lupine
Lupinus arboreus

Bright yellow, or sometimes purple flower; on sandy flats near coast

Mock Locust
Amorpha hispidula
Fragrant; black-purple flower

SUMAC FAMILY *Anacardiaceæ*

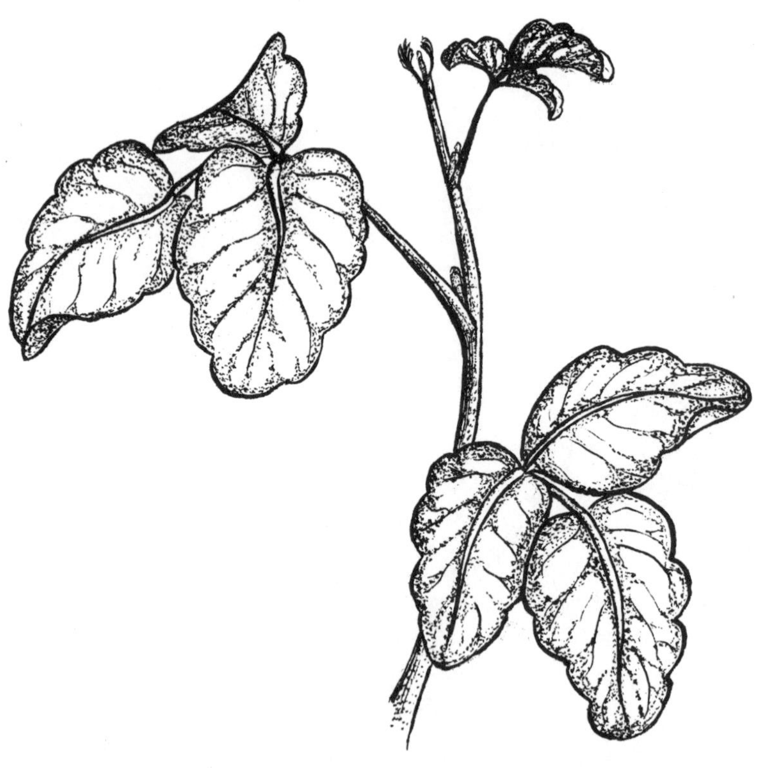

POISON OAK
Rhus diversiloba

Beware!
Three leaflets, white berries; deciduous but oil in the bare branches in winter can still produce an irritating rash. This is not a plant print—the author is allergic to Poison Oak— but a drawing compassionately made by Maggie Cavagnaro

BURNING BUSH FAMILY *Celastraceæ*

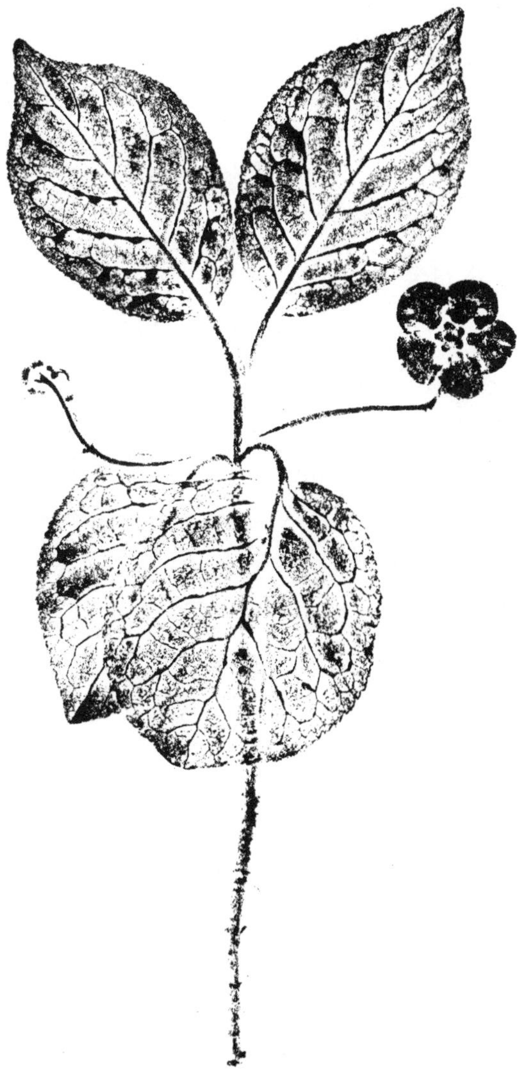

WESTERN BURNING BUSH
Euonymus occidentalis

Maroon flowers, deciduous, on shaded slopes in forests

BUCKTHORN FAMILY *Rhamnaceæ*
A family of trees or shrubs with simple leaves and small flowers in clusters

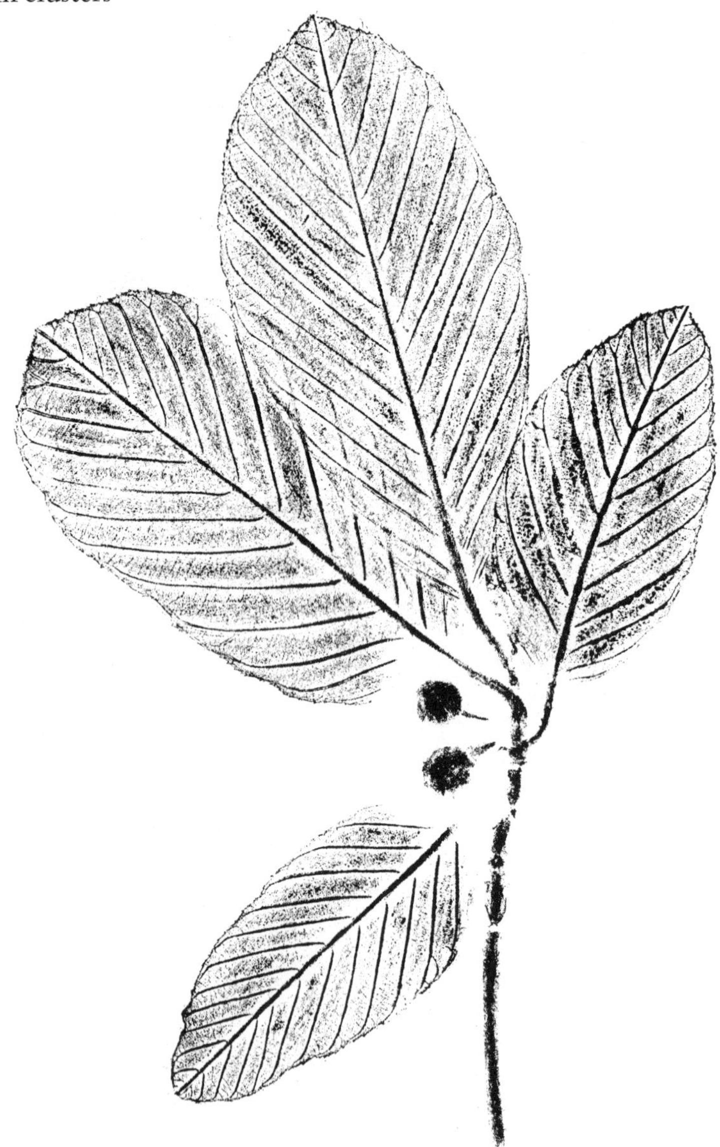

Cascara sagrada, or *Rhamnus purshiana*

Deciduous; Mendocino north; greenish flower

Coffee Berry
Rhamnus californica

Evergreen; on coastal bluffs; yellow flower in December

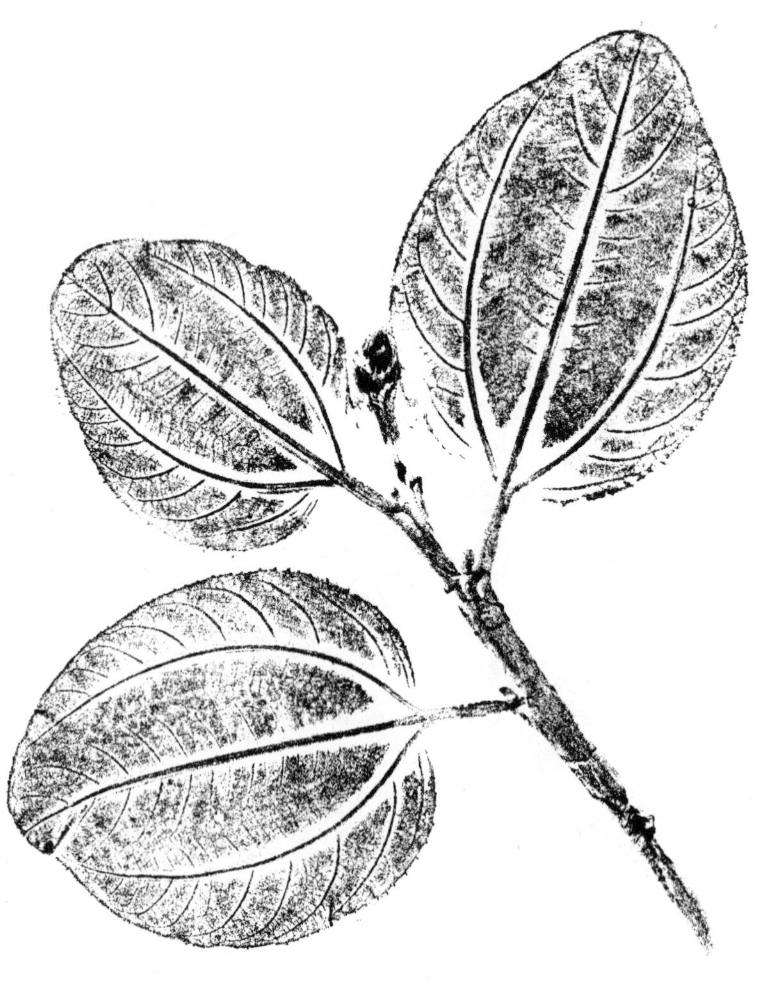

Tobacco Brush
Ceanothus velutinus

Large leaves, extremely waxy; dark seeds; near coast

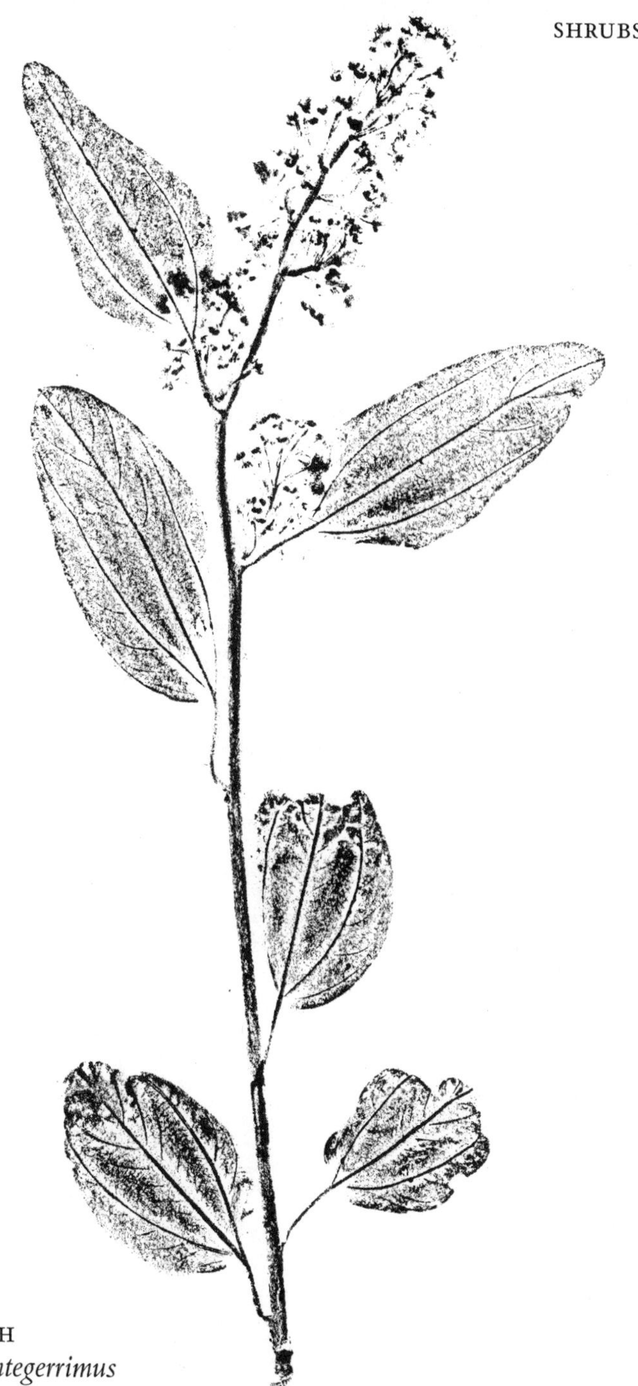

Deer Brush
Ceanothus integerrimus
White flower; deciduous; Sierra Nevada

INDIGO BRUSH
Ceanothus foliosus

Deep indigo flower, low plant in chaparral

Blue Blossom; Wild Lilac
ceanothus thyrsiflorus

Blue flower, tree-like, associated with coast redwoods

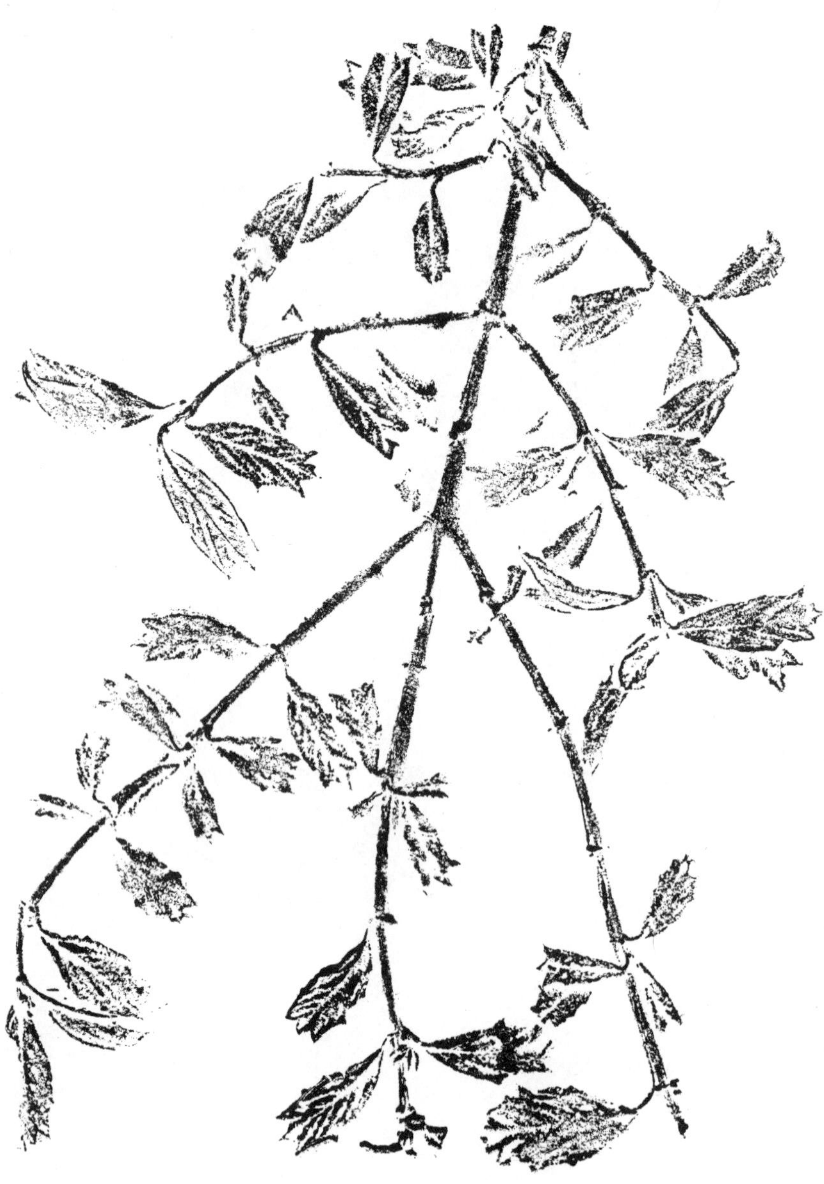

Squaw Mat Carpet; Mahala Mat
Ceanothus prostratus

Grows flat on ground, Sierra Nevada; blue flower

GRAPE FAMILY *vitaceæ*

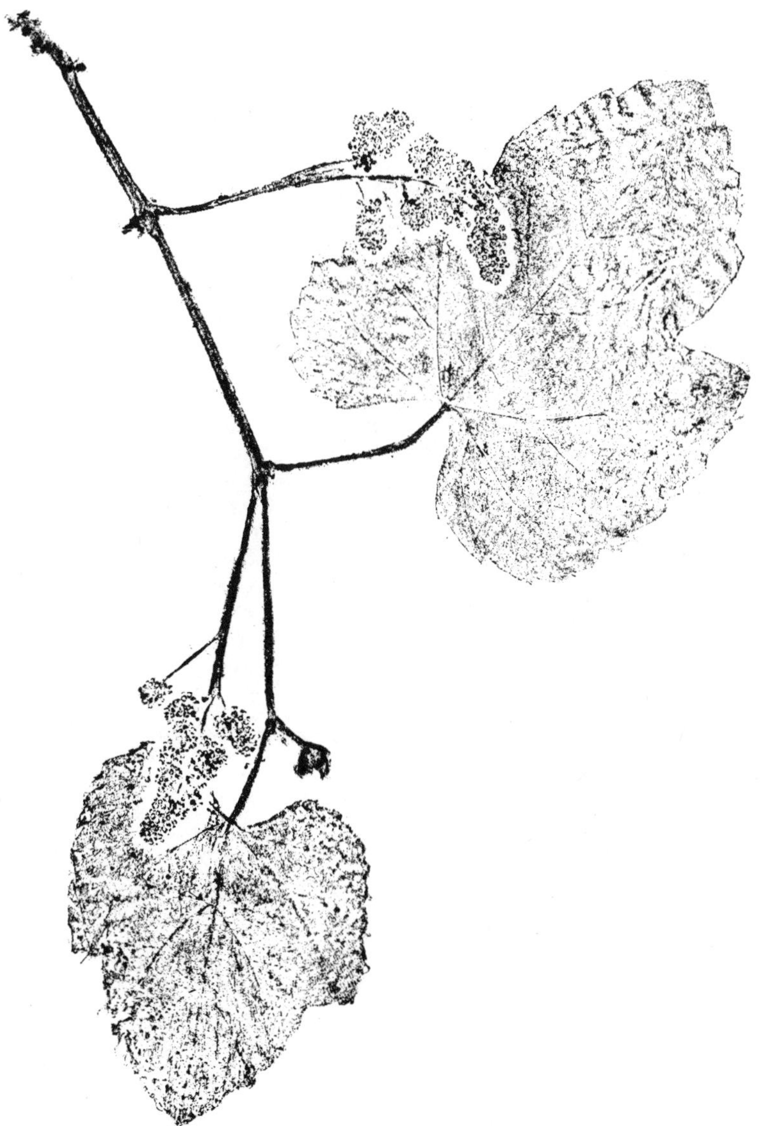

WILD GRAPE
vitis californica

Greenish flower, small purple edible grapes

SILKTASSEL FAMILY *Garryaceæ*

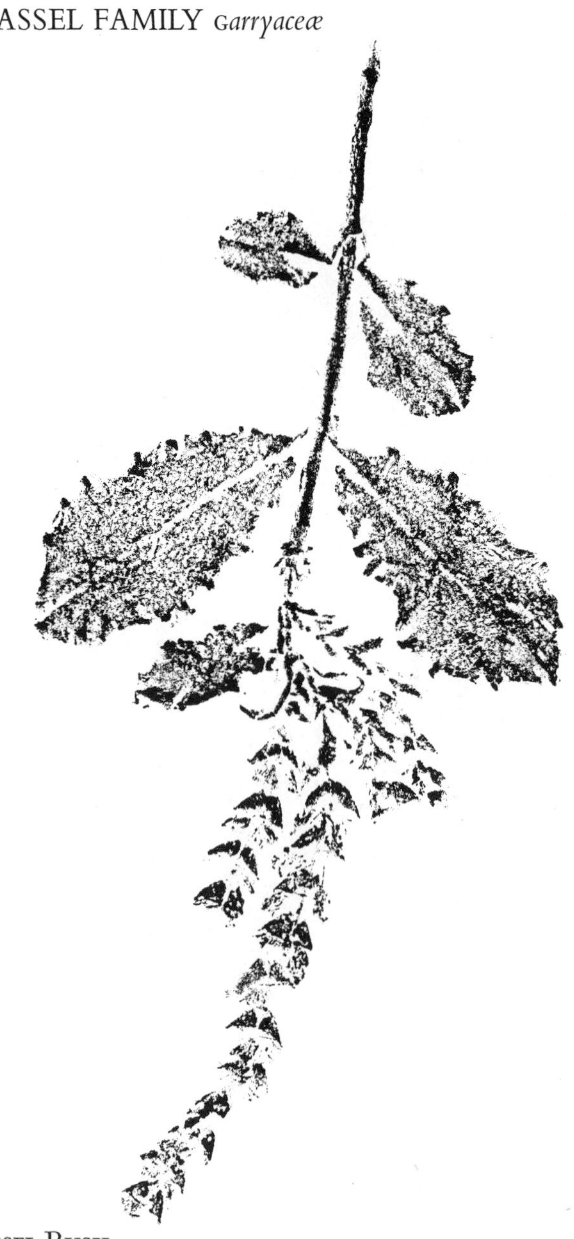

Silktassel Bush
Garrya elliptica

Silk tassels are the male flower

HEATH FAMILY *Ericaceæ*

Includes madrone, manzanitas, Labrador tea, rhododendrons, salal and huckleberry. Flowers are always pink and bell-shaped

BEARBERRY; KINNIKINNIK
Arctostaphylos uva-ursi

Mat-like manzanita also called Sandberry, usually near coast; pinkish flowers, red berries

Cushing Manzanita
Arctostaphylos glandulosa cushingiana

Pink flowers in December, fruit looks like little apples. Chaparral.

Pygmy Forest Manzanita
Arctostaphylos nummularia

Dwarf bush of Mendocino County

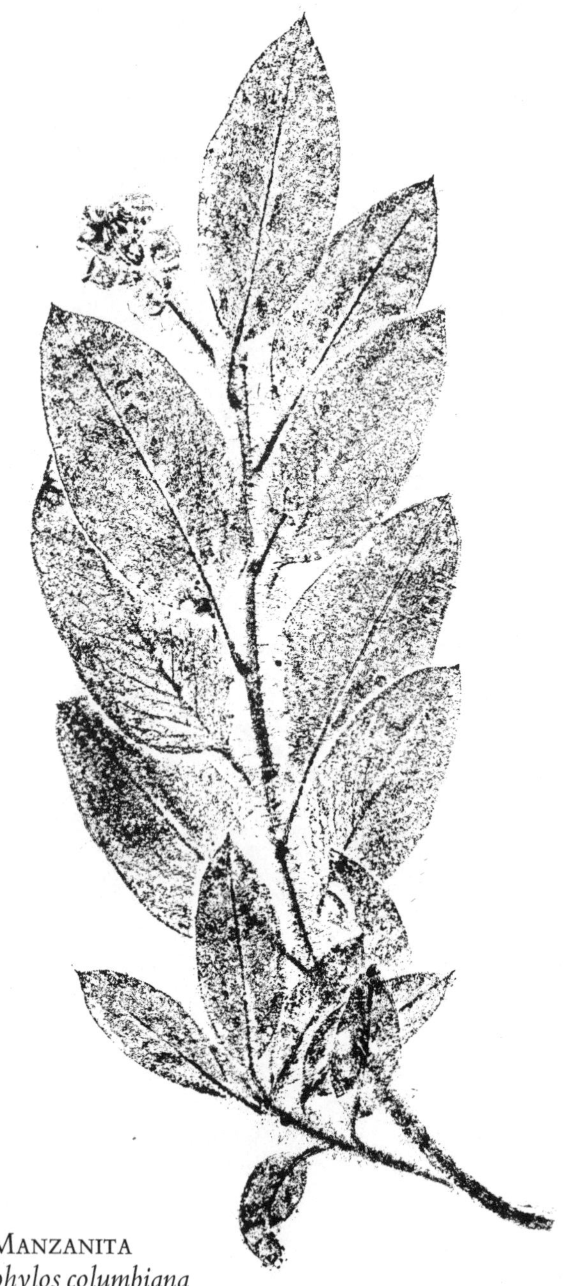

Hairy Manzanita
Arctostaphylos columbiana

Along coast; pale pink flowers in November

Shatterberry Manzanita
Arctostaphylos sensitiva

Shiny heart-shaped leaves, pink flowers in December

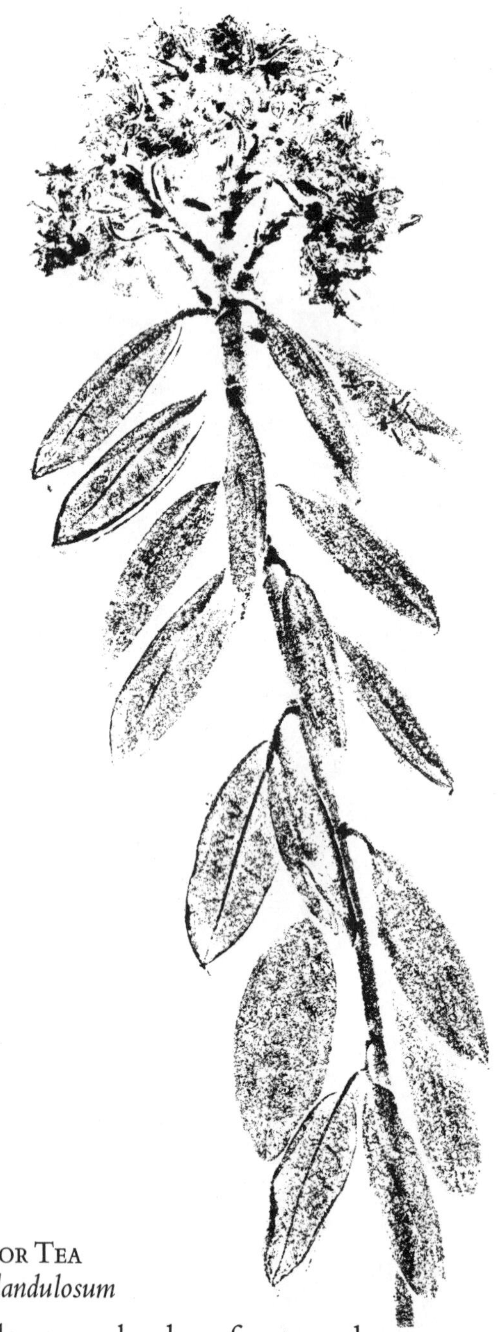

Labrador Tea
Ledum glandulosum
White flower; on borders of wet meadows or marshy places

WESTERN RHODODENDRON
Rhododendron macrophyllum

Large purple flower—buds only shown here

Western Azalea
Rhododendron occidentale

Large white fragrant flower; stream banks and moist places

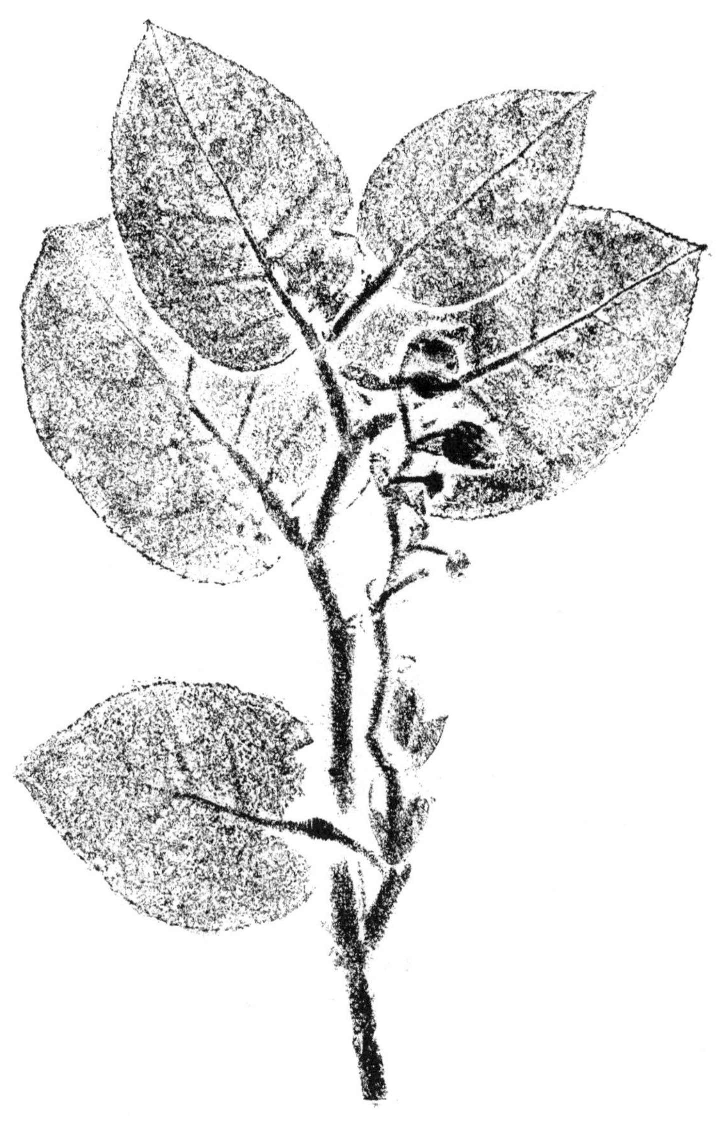

SALAL
Gaultheria shallon

Pink bell-like flowers; associated with coast redwoods

RED HUCKLEBERRY
Vaccinium parvifolium

Thin shrub in shade in coast redwood forest; red berries; deciduous

HUCKLEBERRY
Vaccinium ovatum

Robust shrub, pink bell-like flowers, dark berries. Often dominant shrub in coast redwood forests.

WATERLEAF OR PHACELIA FAMILY *Hydrophyllaceæ*

Leaves usually hairy, divided flowers white to violet

YERBA SANTA
Eriodictyon californicum

Lavender flower, leaves sticky with resin

MINT FAMILY *Labiatæ*

Flowers 2-lipped, stems usually square, leaves opposite, mint odor

Black Sage
salvia mellifera

Lavender flowers in clusters

Pitcher Sage
Lepechinia calycina
Lavender blossoms and strong odor; chaparral and brushy slopes

FIGWORT FAMILY *scrophulariaceæ*
Irregular flowers, many species in California

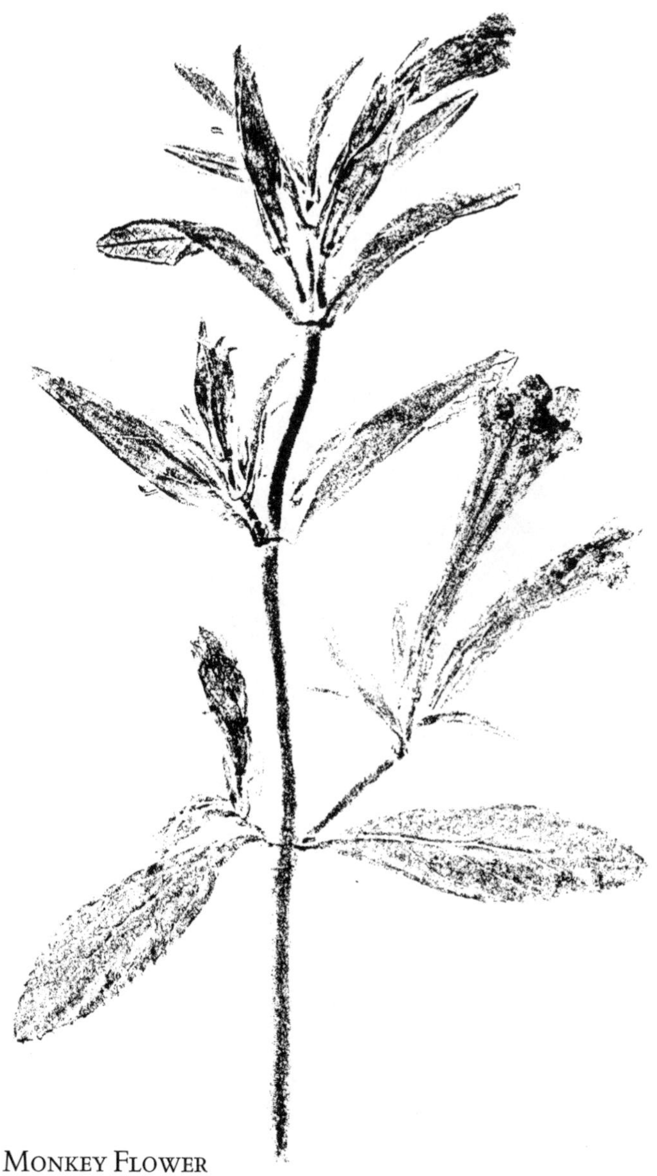

BUSH MONKEY FLOWER
Mimulus aurantiacus

Orange flower; plants bloom 11 months out of the year

HONEYSUCKLE FAMILY *caprifoliaceæ*

Flowers often in axil of leaf, leaves simple and opposite

RED ELDERBERRY
Sambucus callicarpa

Grows near coast; creamy flowers in spire-shape clusters, red berries

Blue Elderberry
sambucus cœrulea

Sometimes tree-like; creamy flowers in flat-topped clusters, blue berries edible

Snowberry
symphoricarpos mollis

Pink flowers, then white berries in winter

TWINBERRY
Lonicera ledebourii

Yellow flower, dark berries

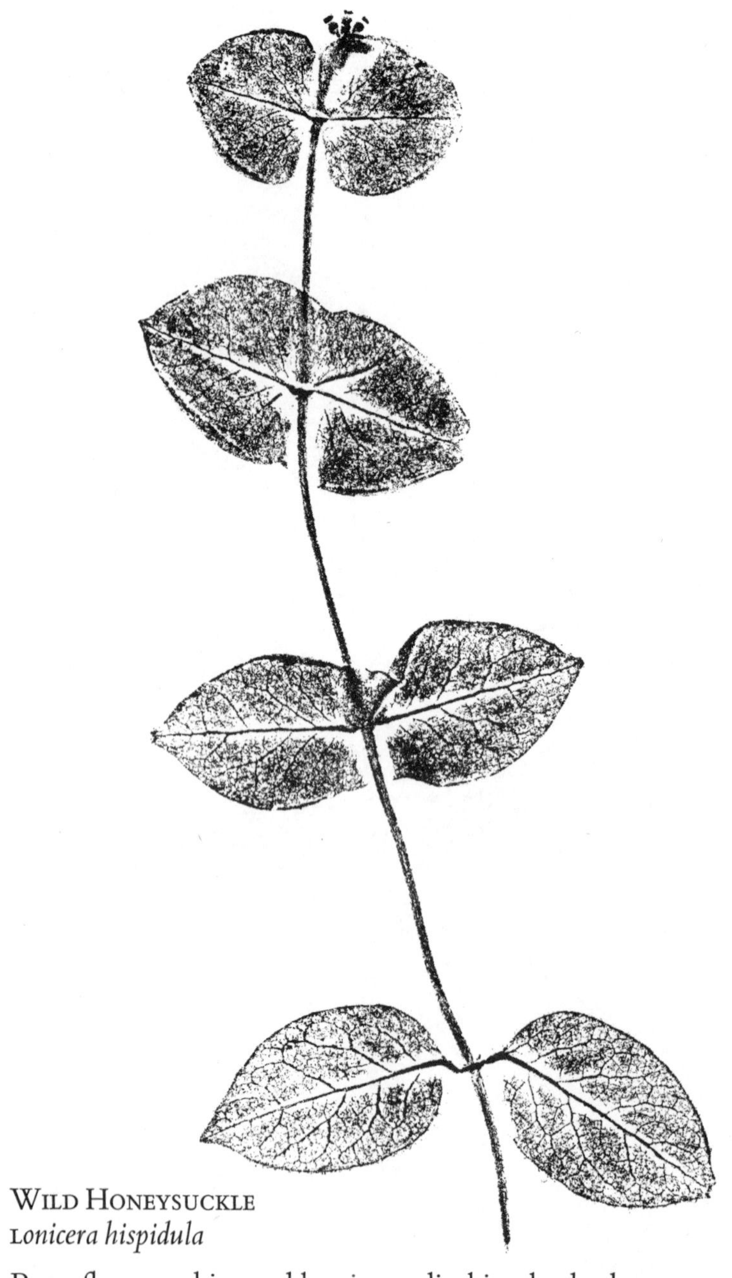

Wild Honeysuckle
Lonicera hispidula

Rosy flowers, shiny red berries; a climbing bush, the uppermost leaves grow around the stem

SUNFLOWER FAMILY *compositæ*

MOCK HEATHER
Haplopappus ericoides

Yellow flower; low-growing stiff shrub on coastal slopes

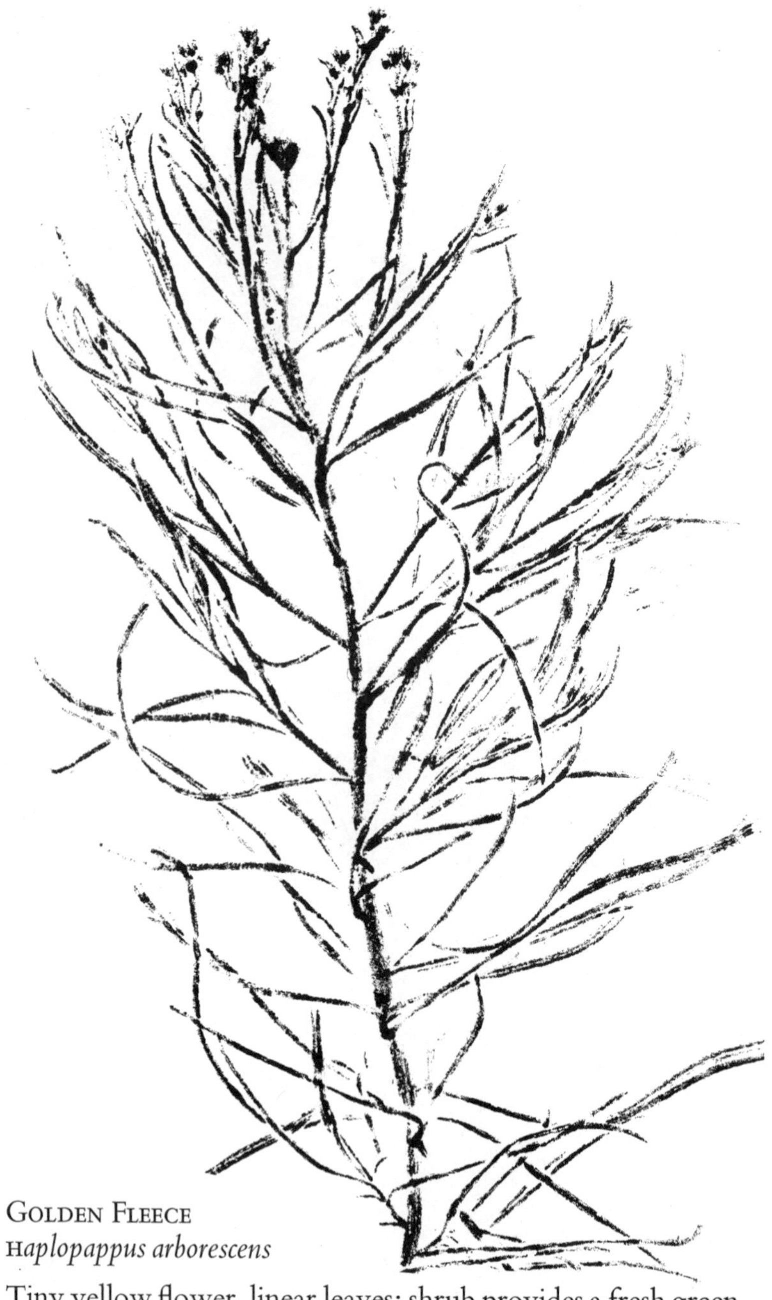

Golden Fleece
Haplopappus arborescens

Tiny yellow flower, linear leaves; shrub provides a fresh green color in the chaparral in late summer

RABBIT BRUSH
Chrysothamnus nauseosus

Yellow flowers in fall, Sierra Nevada

Coyote Bush
Baccharis pilularis

White fuzzy flowers in the fall; also called Chaparral Broom

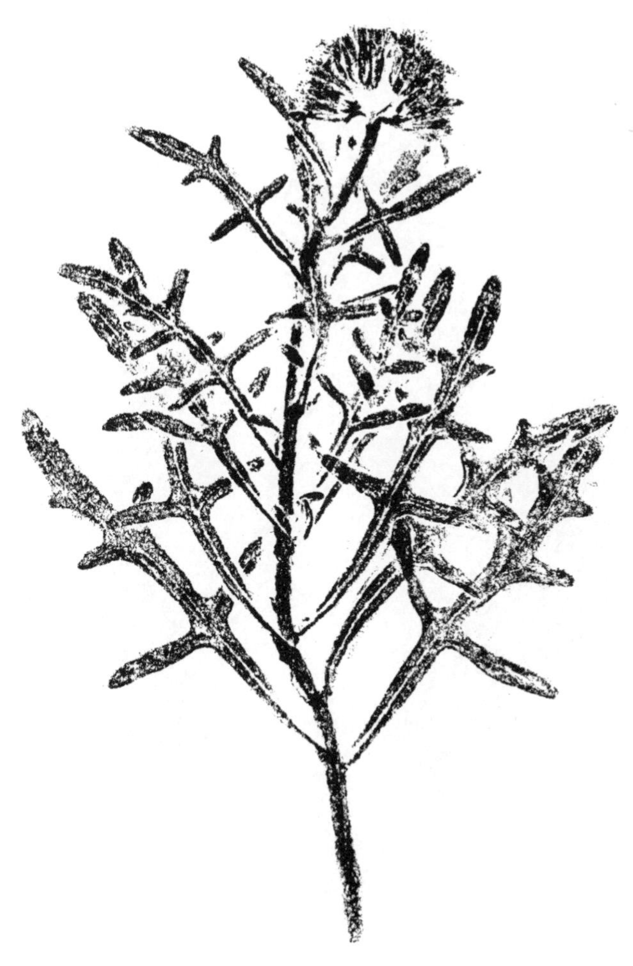

LIZARD TAIL [TARWEED TRIBE]
Eriophyllum stæchadifolium

At the beach, round-headed yellow flowers on low rounded bushes

COMMON SAGEBRUSH [MAYWEED TRIBE]
Artemisia tridentata

Widespread shrub on east slope of Sierra Nevada; grayish leaves, flowers in panicles in late summer

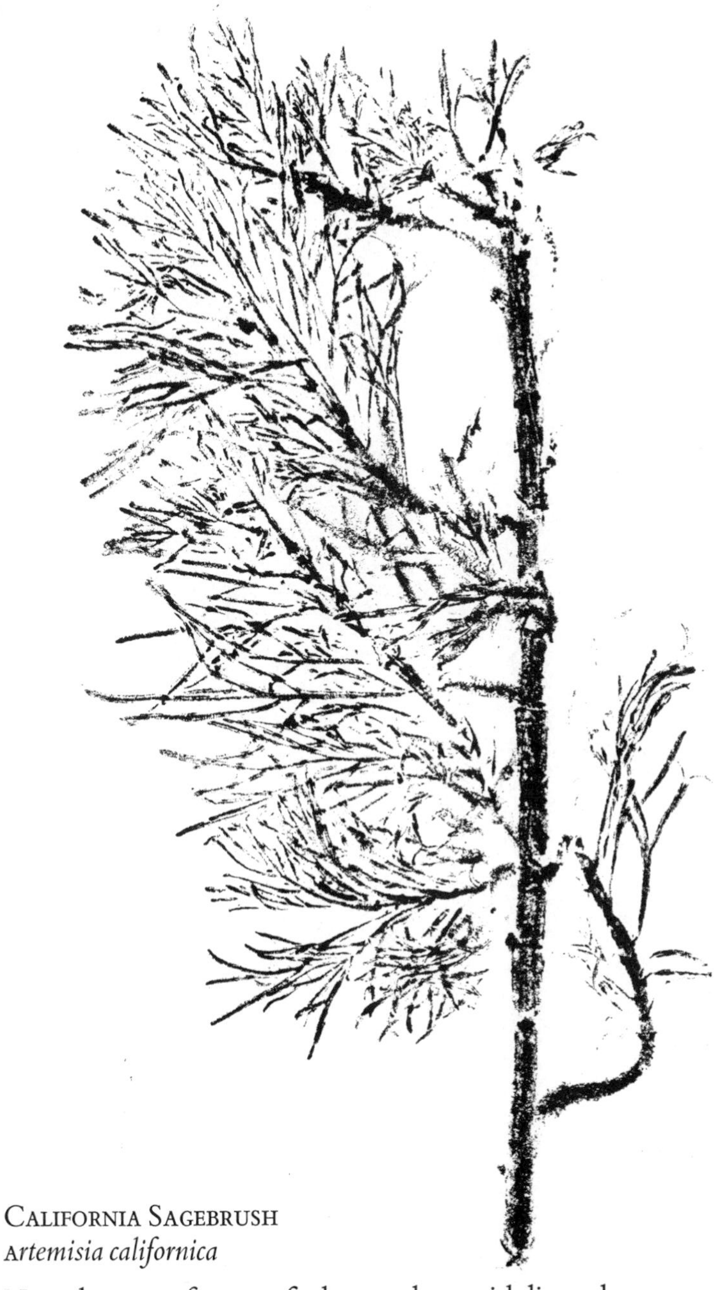

California Sagebrush
Artemisia californica

Near the coast, forms soft chaparral; grayish linear leaves

NOTES

Trees

TREES are defined as woody plants with a tall trunk. Their flowers are often overlooked because they are high.

CONIFERS (*Gymnosperms*) are evergreen trees with needle-like or scale-like foliage, usually cone-bearing, with naked seeds borne in the cones.

BROADLEAF trees (*Angiosperms*) can be evergreen or deciduous, with seeds enclosed within a fruit.

Conifer Trees *Gymnosperms*

YEW FAMILY *Taxaceæ*

CALIFORNIA NUTMEG
Torreya californica

Needles have sharp pointed ends

WESTERN YEW
Taxus brevifolia

Short green needles and round red fruit; northwest coast, uncommon

PINE FAMILY *Pinaceæ*
(includes pines, larches, spruces, hemlocks, true firs, Douglas-fir)

WHITE PINE
Pinus monticola

Needles in 5's; Sierra Nevada

Sugar Pine
Pinus lambertiana

Needles in 5's; Sierra Nevada; extremely large cones

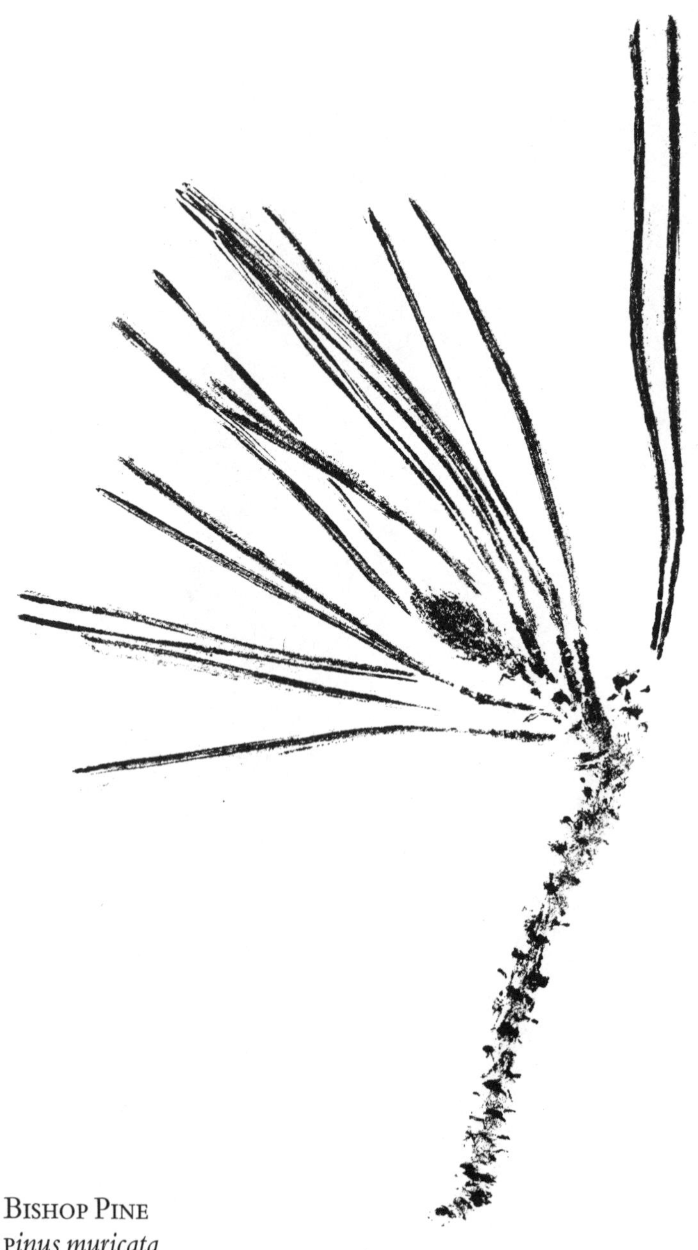

Bishop Pine
Pinus muricata

Needles in 2's; grows along the coast; cones open after fires

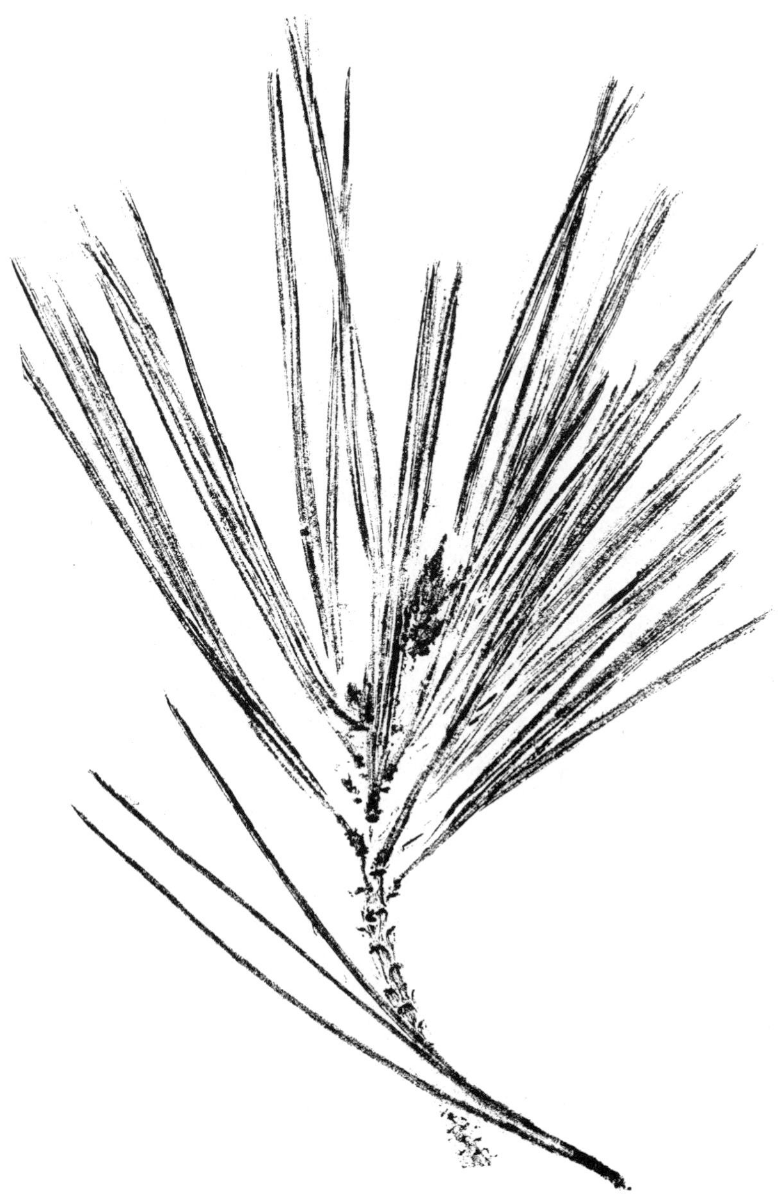

PONDEROSA PINE; YELLOW PINE
Pinus ponderosa

Needles in 3's; Sierra Nevada and Coast Ranges

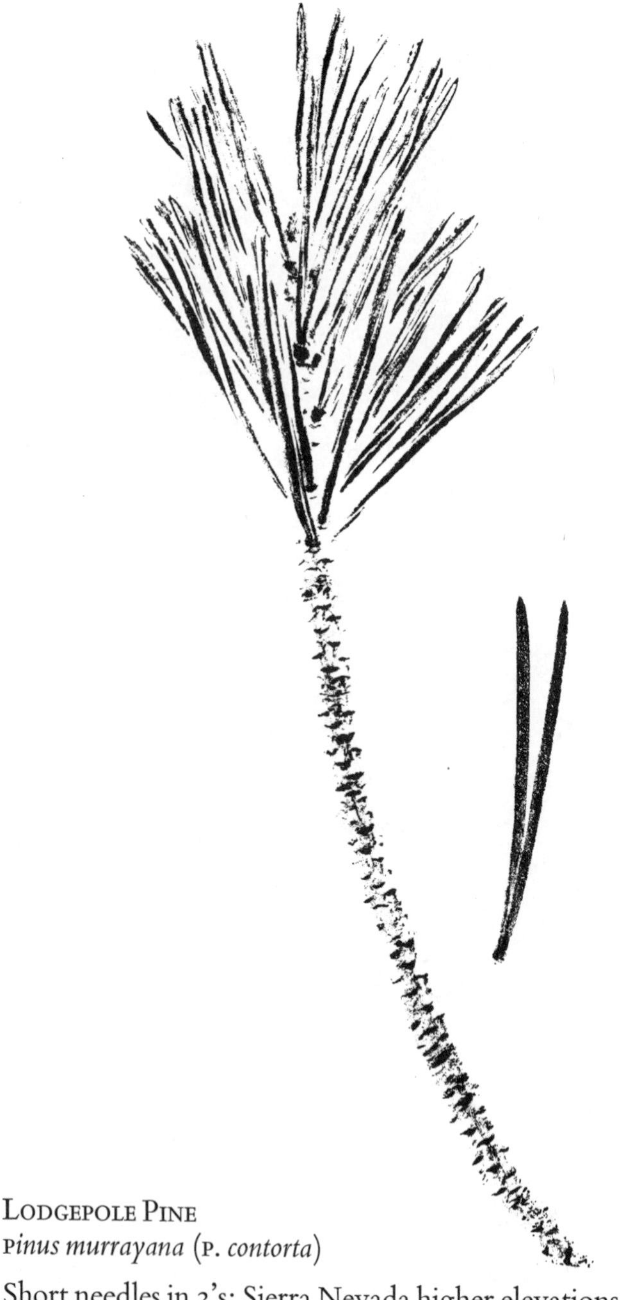

LODGEPOLE PINE
Pinus murrayana (P. *contorta*)
Short needles in 2's; Sierra Nevada higher elevations

Beach Pine
Pinus contorta

Short needles in 2's; stunted trees along the coast

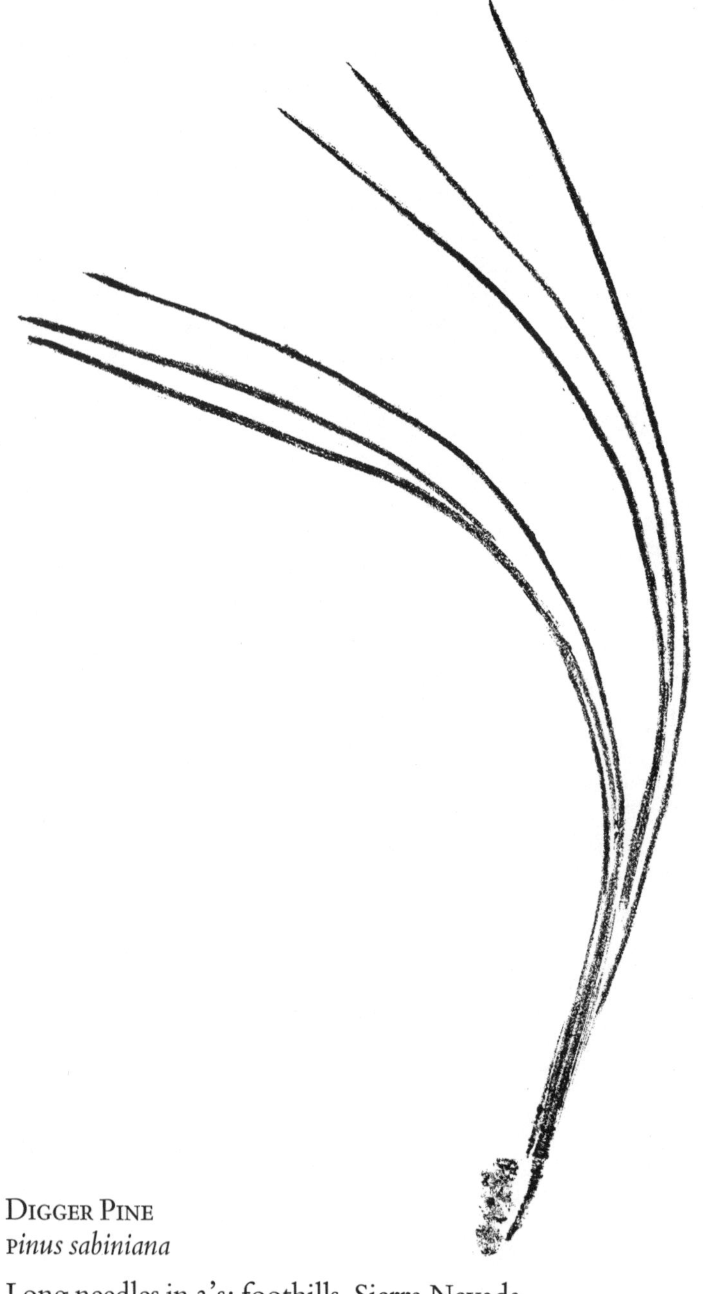

Digger Pine
Pinus sabiniana

Long needles in 3's; foothills, Sierra Nevada

Pinyon Pine
Pinus monophylla

Single needles; Sierra Nevada

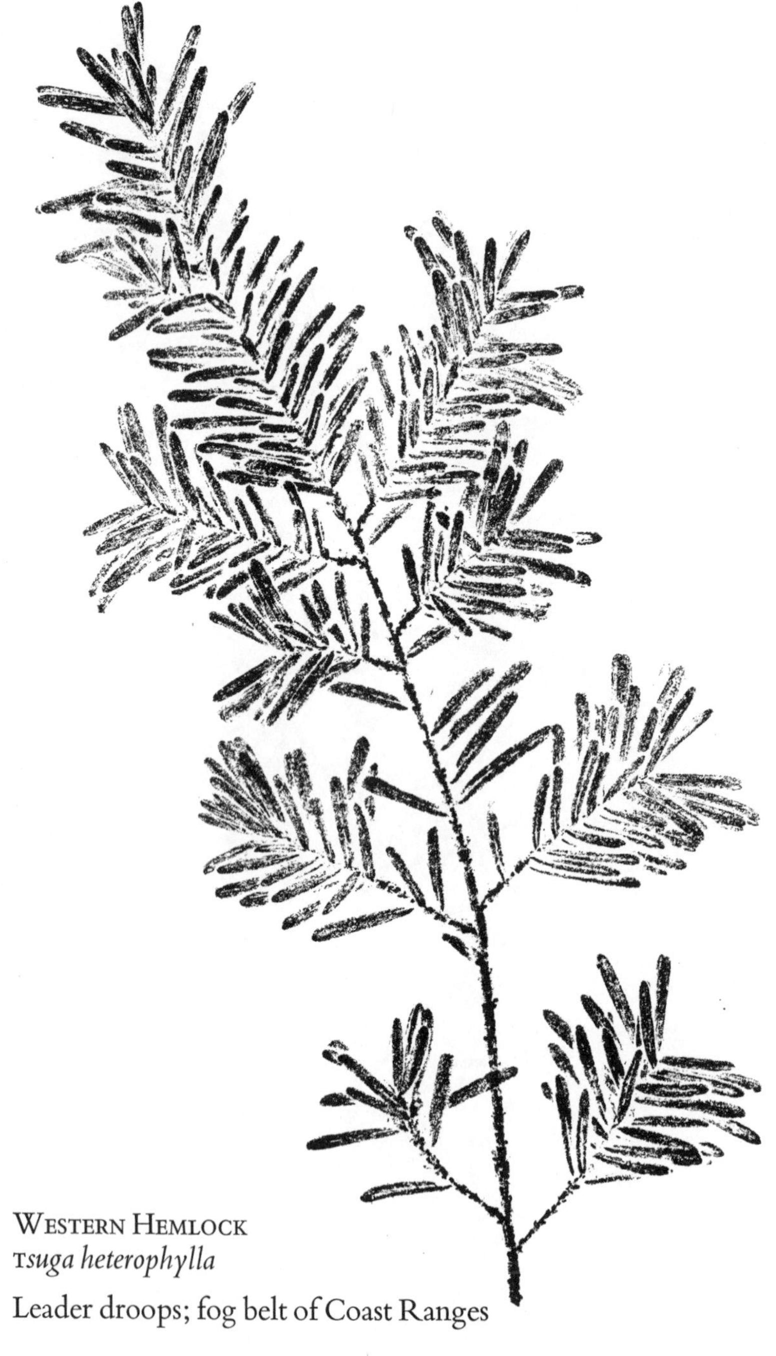

Western Hemlock
Tsuga heterophylla

Leader droops; fog belt of Coast Ranges

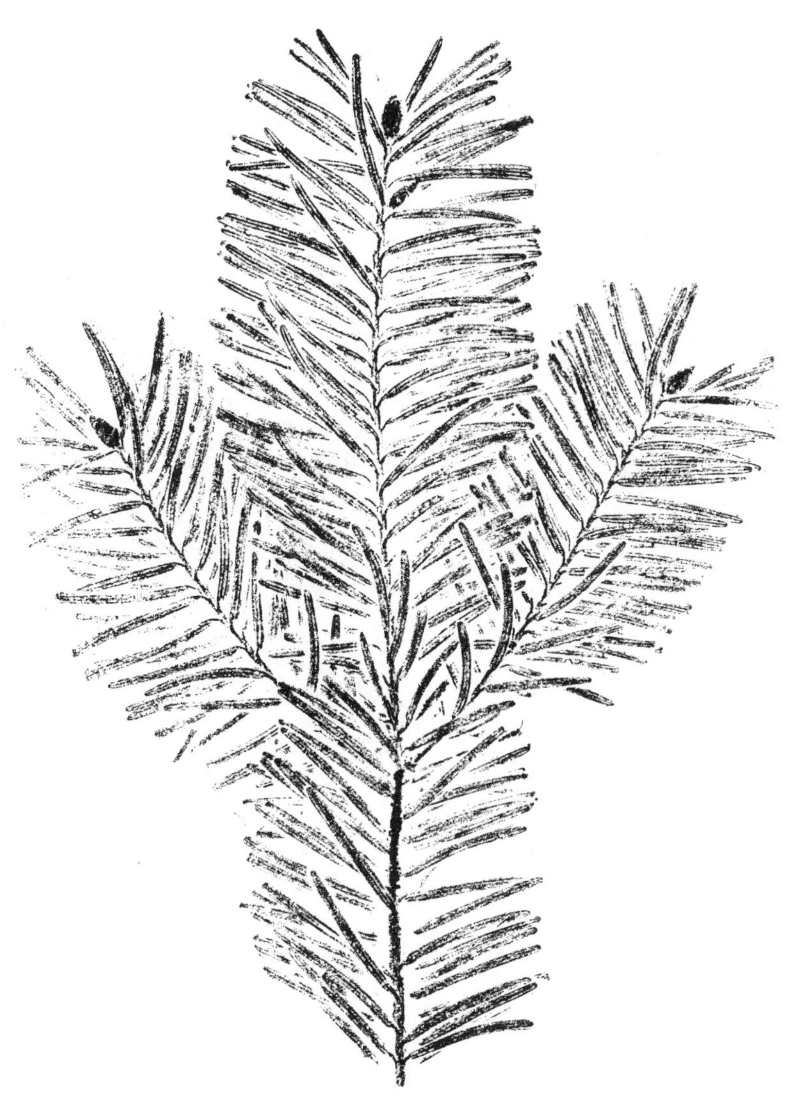

DOUGLAS FIR
Pseudotsuga menziesii

Not a fir, not a hemlock; associated with coast redwoods; also south in Sierra beyond Yosemite region

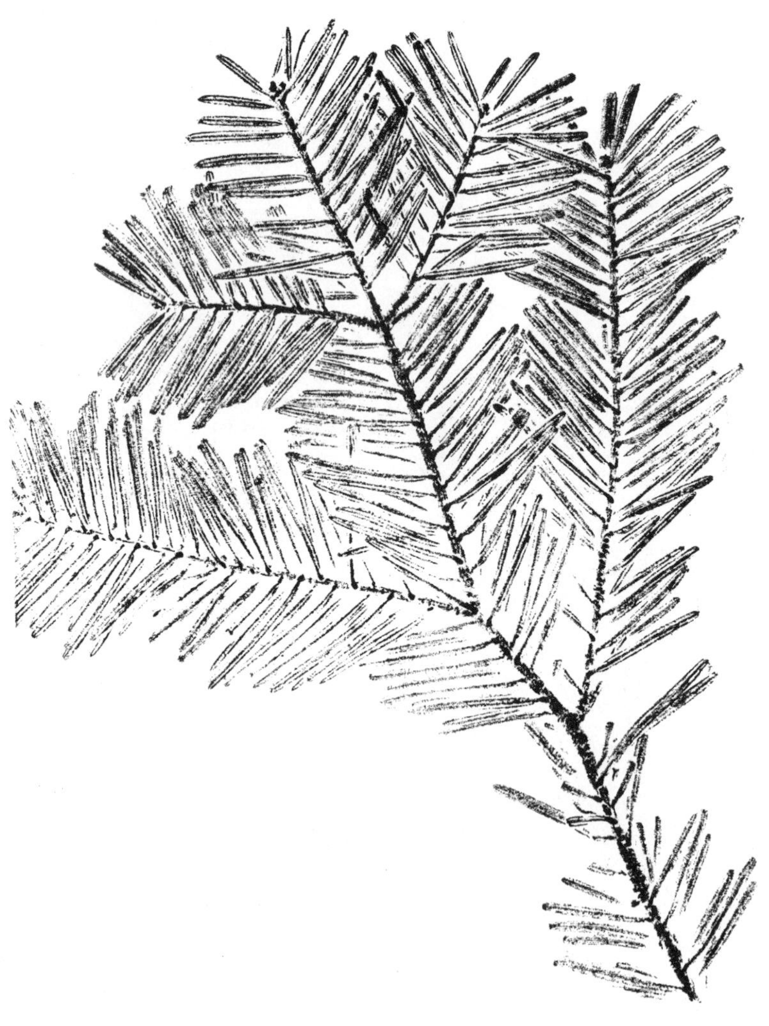

White Fir
Abies concolor

Whitish bark on young trees, brown on older trees, in mountains

Lowland Fir
Abies grandis

Near coast, with coast redwoods in northern part of their range

REDWOOD FAMILY *Taxodiaceæ*

Coast Redwood
Sequoia sempervirens
World's tallest tree

Sierra Redwood (upper left)
Sequoiadendron giganteum
"Big Tree" of the Sierra. Older than Coast Redwood

CEDAR FAMILY *cupressaceæ*
(includes cedars, cypress, junipers)

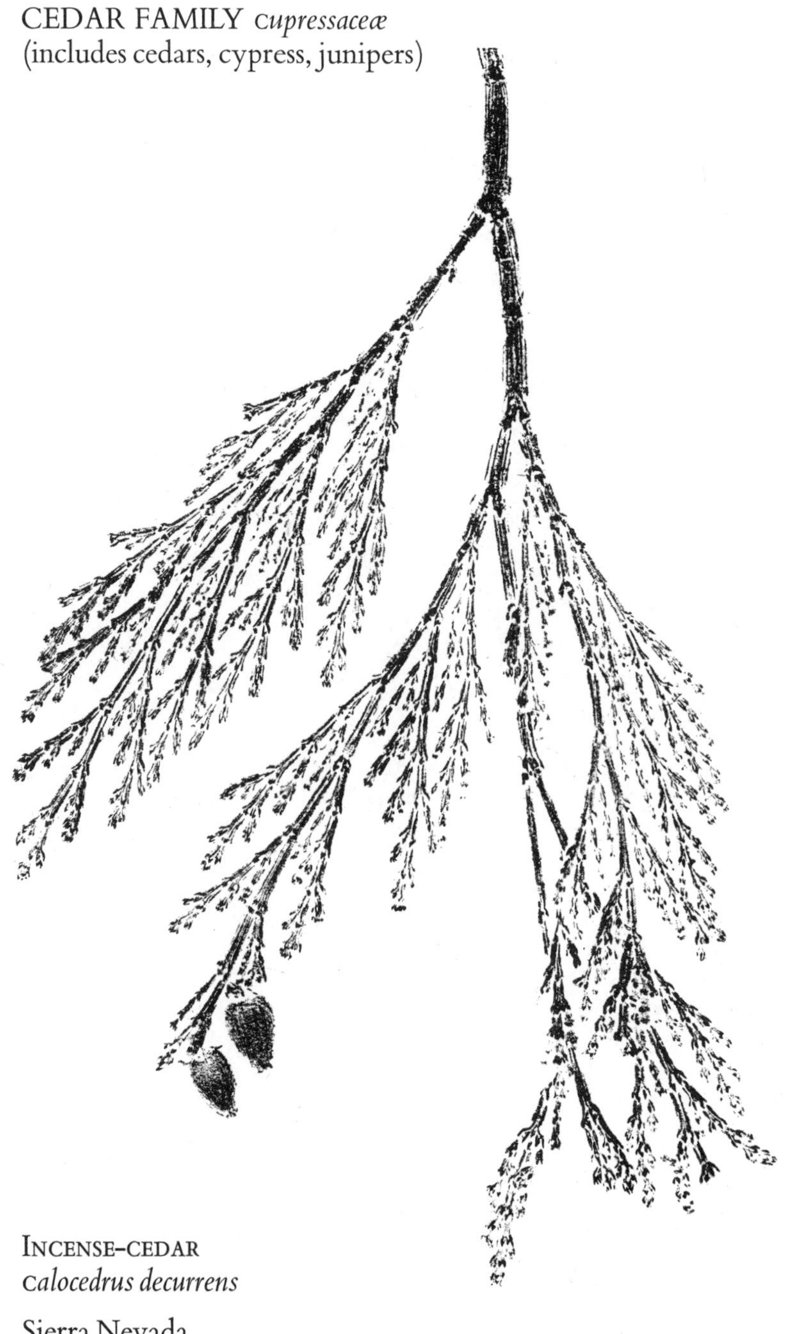

Incense-cedar
calocedrus decurrens

Sierra Nevada

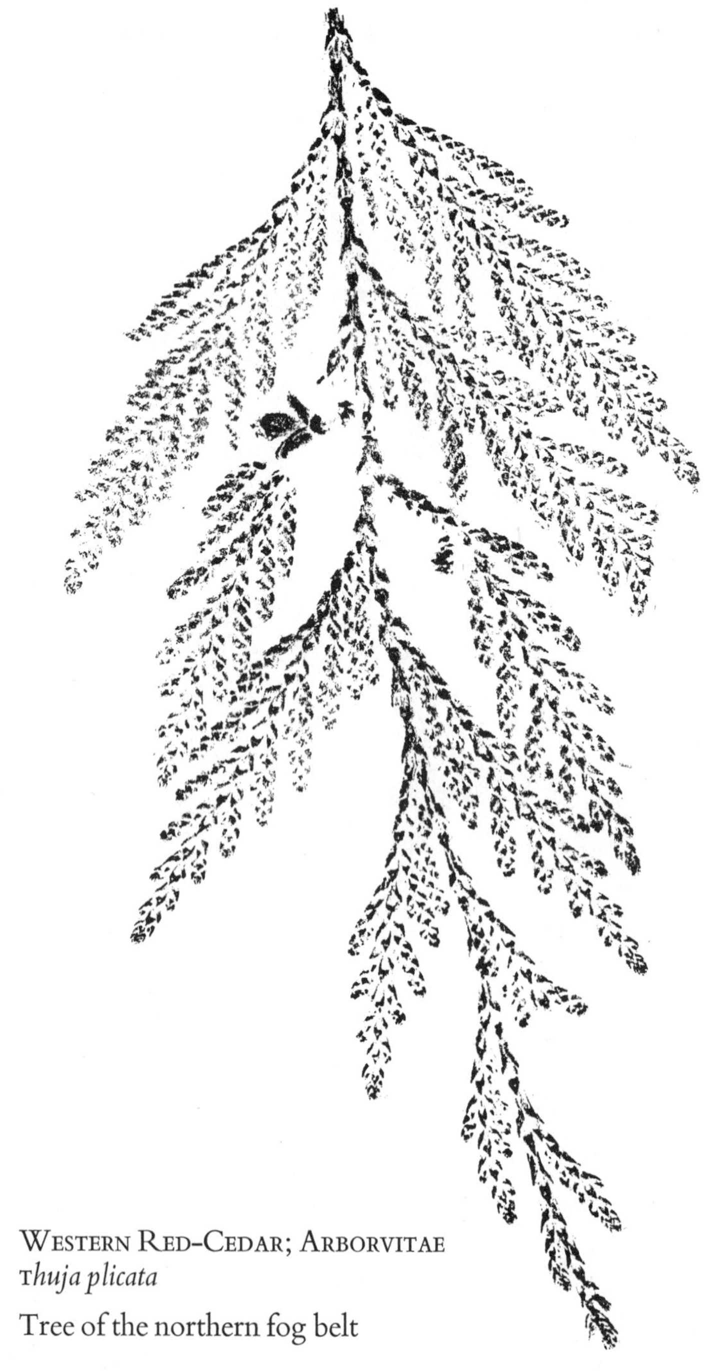

Western Red-Cedar; Arborvitae
Thuja plicata

Tree of the northern fog belt

Lawson Cypress; Port Orford Cedar
Chamæcyparis lawsoniana

An associate of coast redwoods in the north; near coast

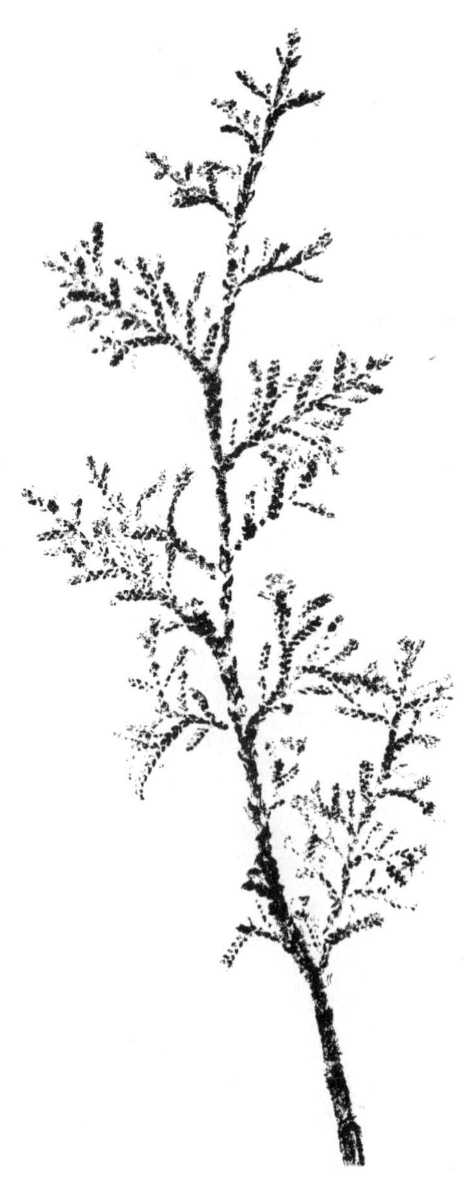

Pygmy Cypress
cupressus pygmaea
Coast of Mendocino County

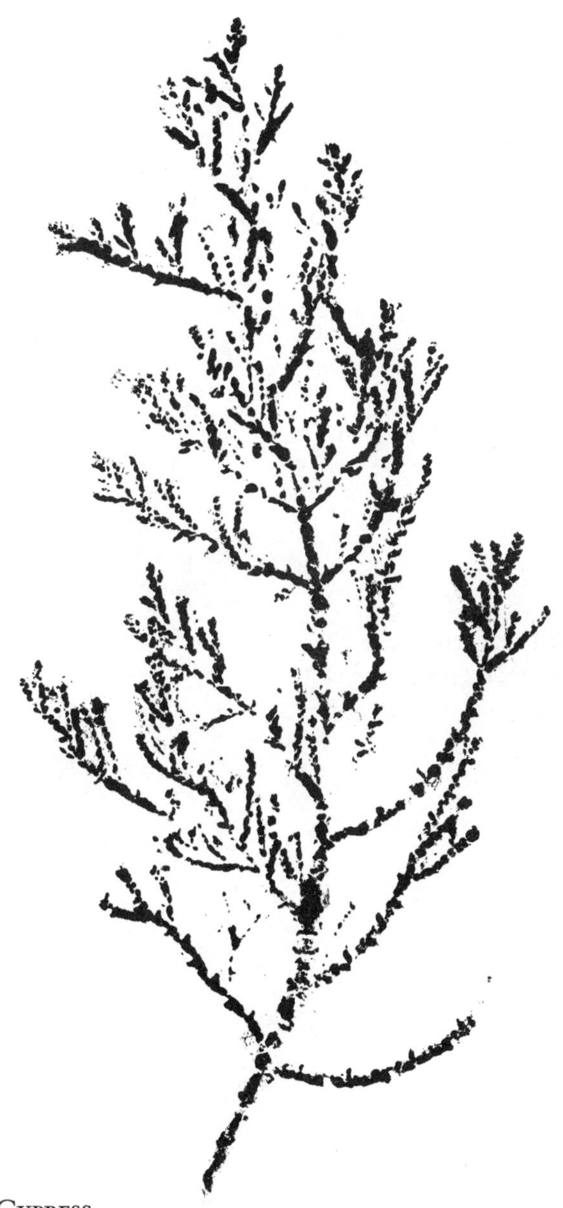

Sargent Cypress
cupressus sargentii

Coast Ranges

Broadleaf Trees *Angiosperms*

Produce flowers, seeds enclosed within a fruit, can be evergreen or deciduous.

WILLOW FAMILY *salicaceæ*
Deciduous and dioecious:
male flowers and female flowers,
in catkins, grow on separate plants.
Often found along streams

SCOULER WILLOW
salix scouleriana

Coast Ranges and Sierra Nevada

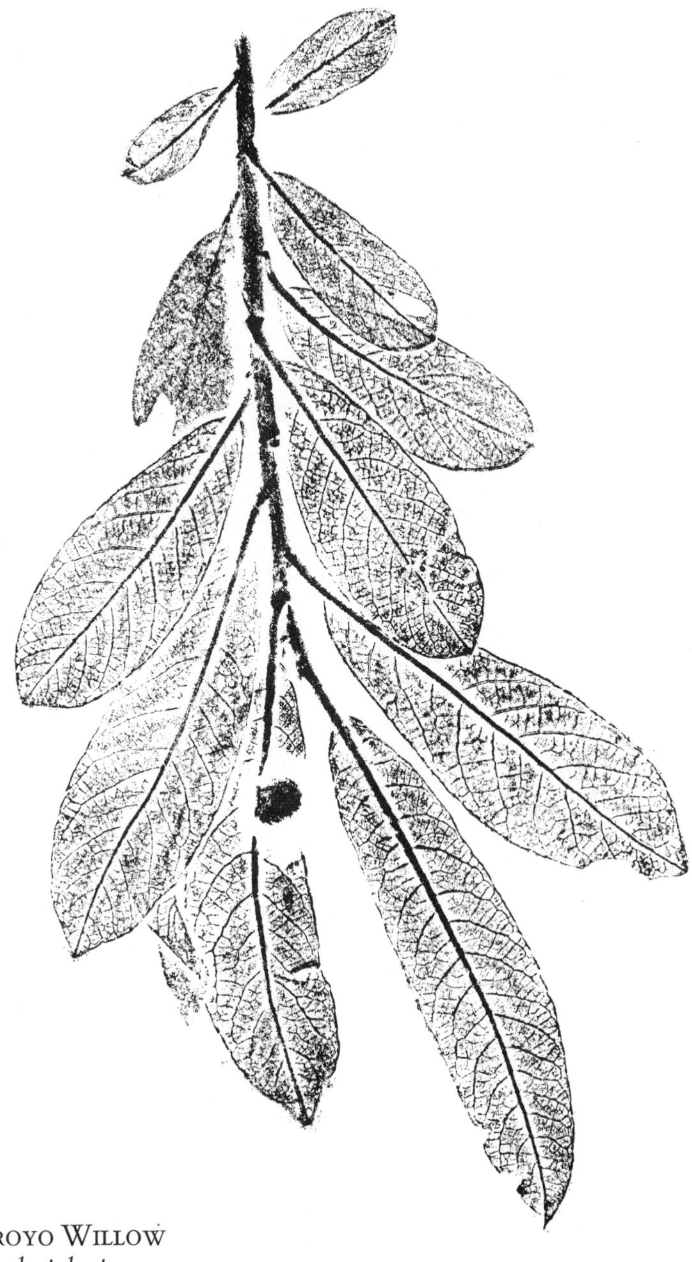

ARROYO WILLOW
salix lasiolepis

Coast Ranges and Sierra Nevada foothills

Black Poplar; Black Cottonwood
Populus trichocarpa

Coast Ranges and Sierra Nevada, higher elevations

FREMONT POPLAR; FREMONT COTTONWOOD
populus fremontii

Inner Coast Ranges, low elevations and Sierra foothills

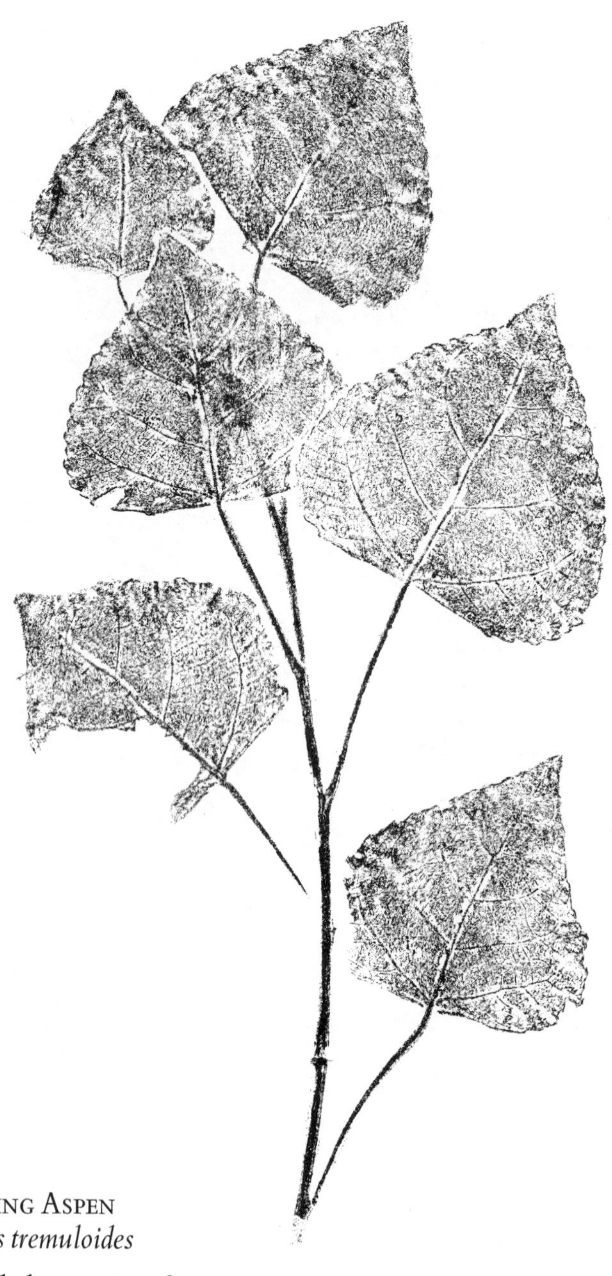

Quaking Aspen
populus tremuloides

Variable leaves, in 3 forms

WAXMYRTLE FAMILY *Myricaceæ*

WAX-MYRTLE; BAYBERRY
Myrica californica

Often shrub-like

BIRCH FAMILY *Betulaceæ*
Deciduous and monoecious: male flowers in catkins, female flowers in catkins or clusters, on the same plant

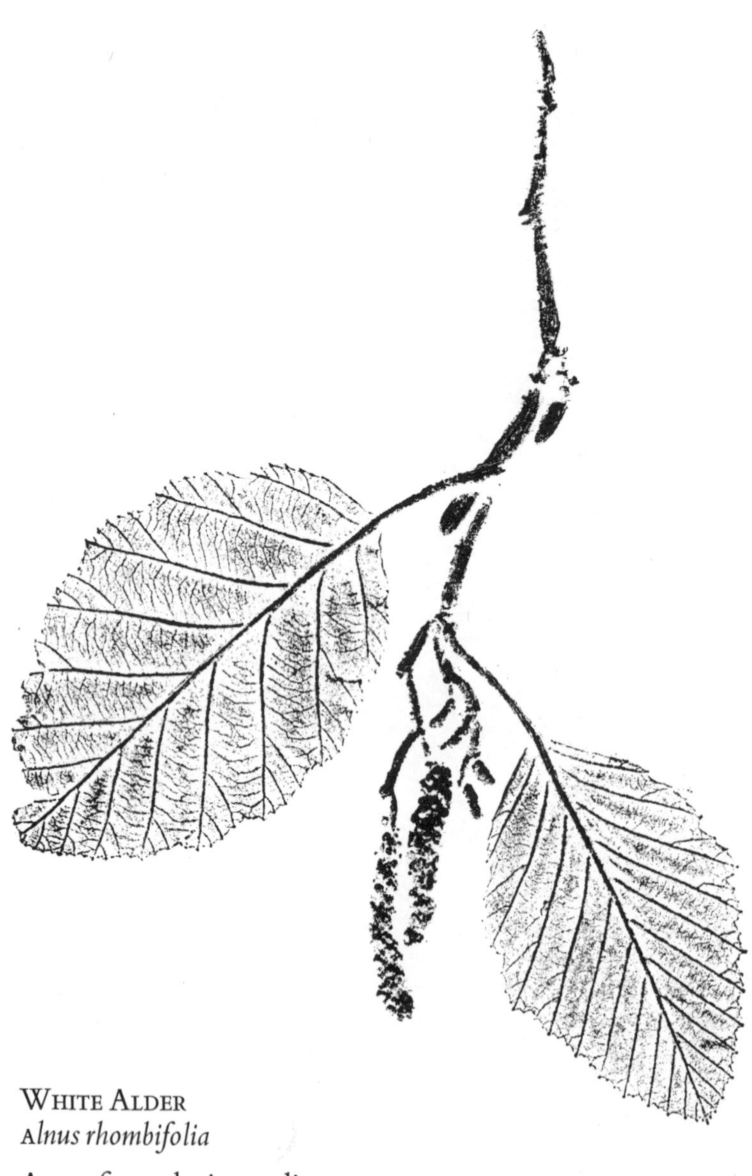

WHITE ALDER
Alnus rhombifolia

Away from the immediate coast

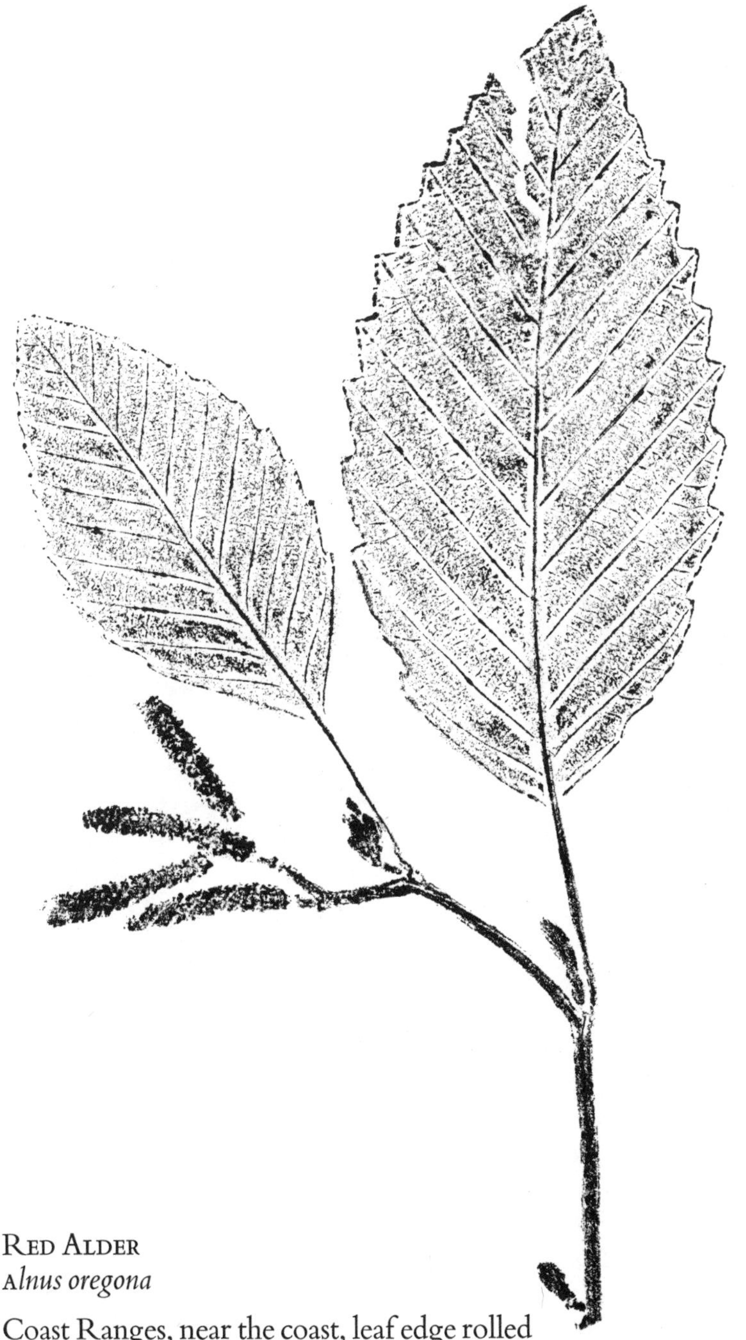

Red Alder
Alnus oregona
Coast Ranges, near the coast, leaf edge rolled

HAZEL FAMILY *corylaceæ*

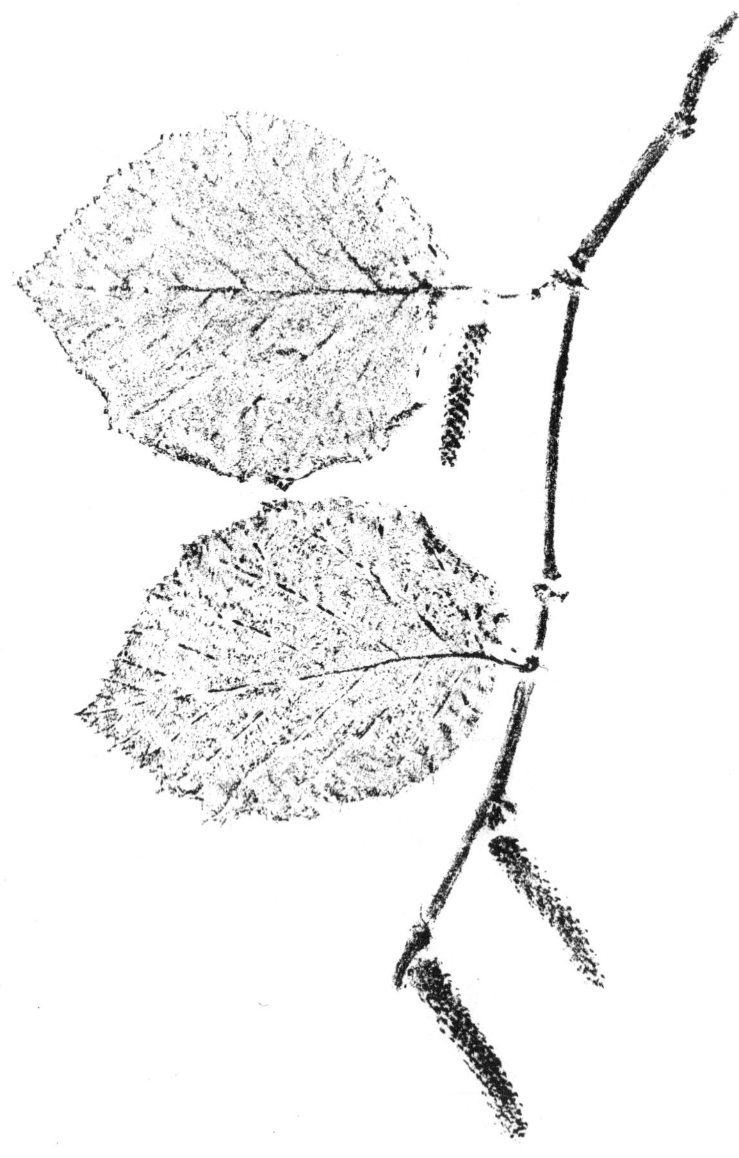

CALIFORNIA HAZEL
corylus californica

A small tree, leaves that feel soft, and edible nut

OAK FAMILY *Fagaceæ*
Fruit a nut or acorn

CHINQUAPIN
Castanopsis chrysophylla

Sometimes shrub-like; underside of leaf bright yellow

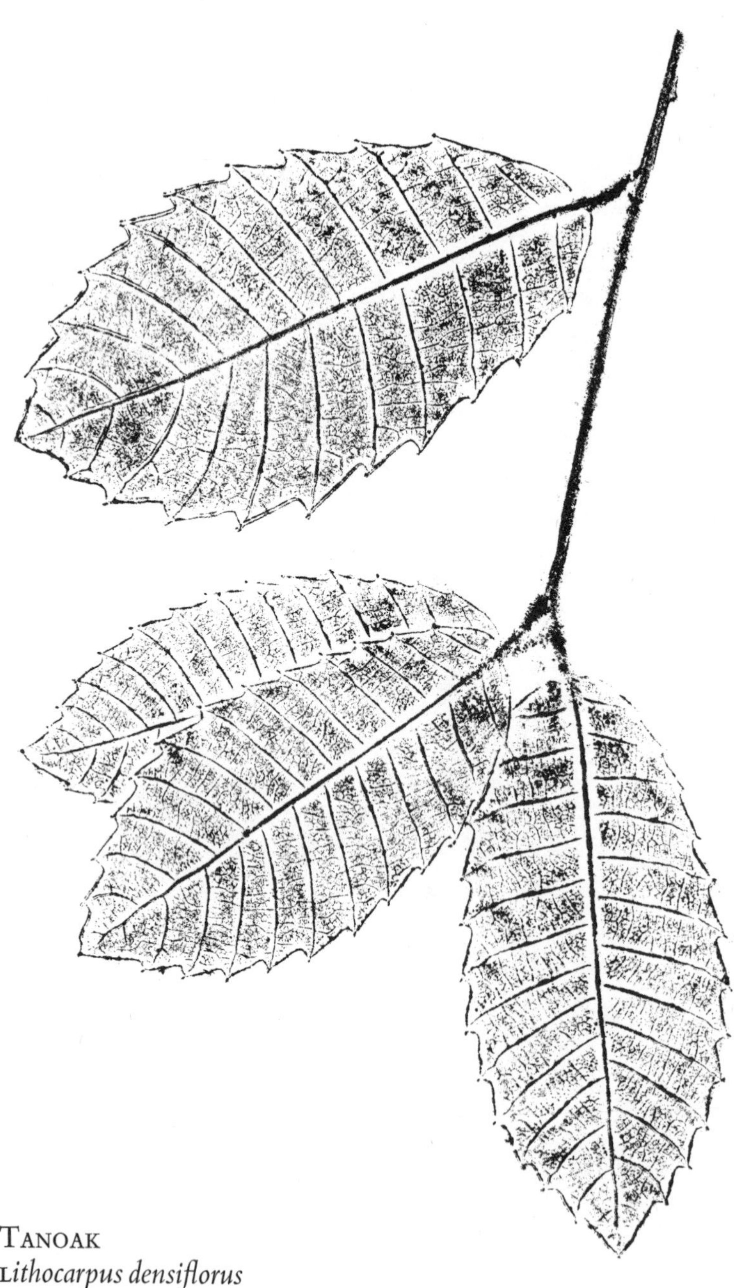

TANOAK
Lithocarpus densiflorus

Associated with coast redwoods

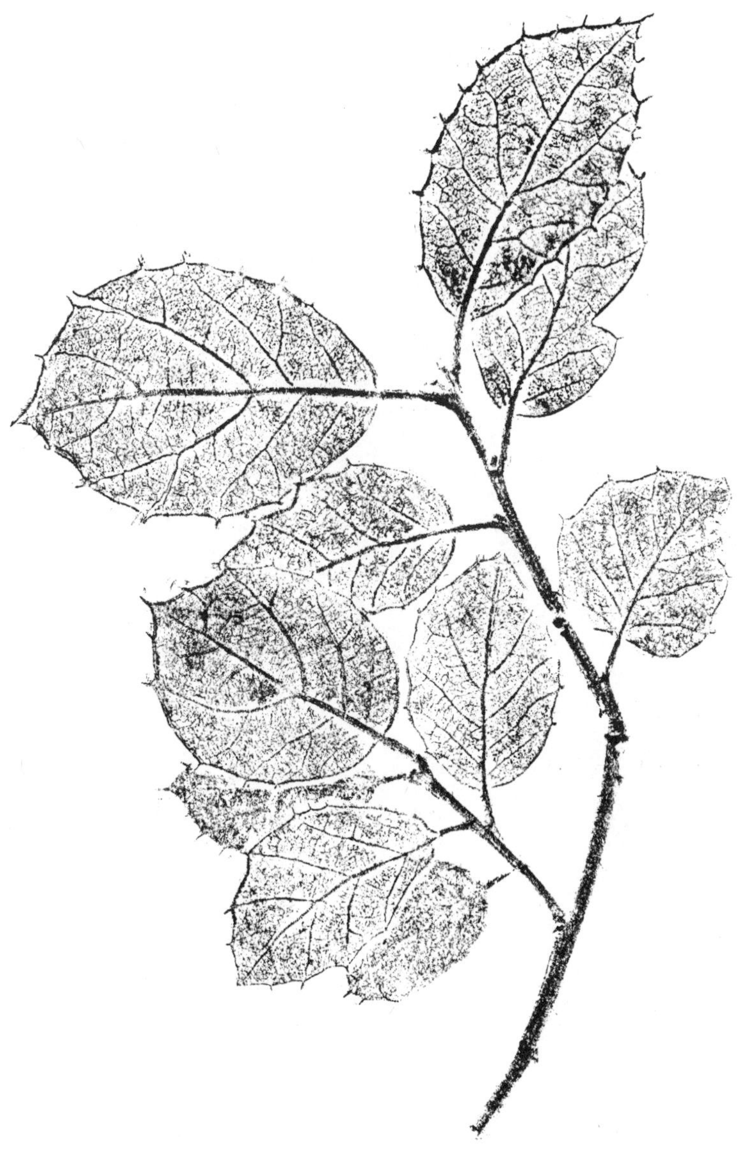

COAST LIVE OAK
Quercus agrifolia

Leaf humped up, no more than 5 veins

CHAPARRAL OAK; INTERIOR OAK
Quercus wislizeni

Leaf flat, 6–10 veins

California Black Oak
Quercus kelloggii

Deciduous, veins end in bristles

Oracle Oak
Quercus morehus

Not a species but a hybrid between Q. *wislizeni* and Q. *kelloggii*; scattered low trees

CANYON LIVE OAK; GOLDCUP OAK
Quercus chrysolepis

Two kinds of leaves, toothed and entire, undersides golden or bluish-gray

Garry Oak; Oregon Oak
Quercus garryana

Deciduous; broad leaf, small round acorn; small tree in shady canyons

VALLEY OAK
Quercus lobata

Deciduous; symmetrical leaf; long large acorn, large tree on open valley grasslands

Blue Oak
Quercus douglasii

Deciduous; leaves have bluish tinge

LAUREL FAMILY *Lauraceæ*

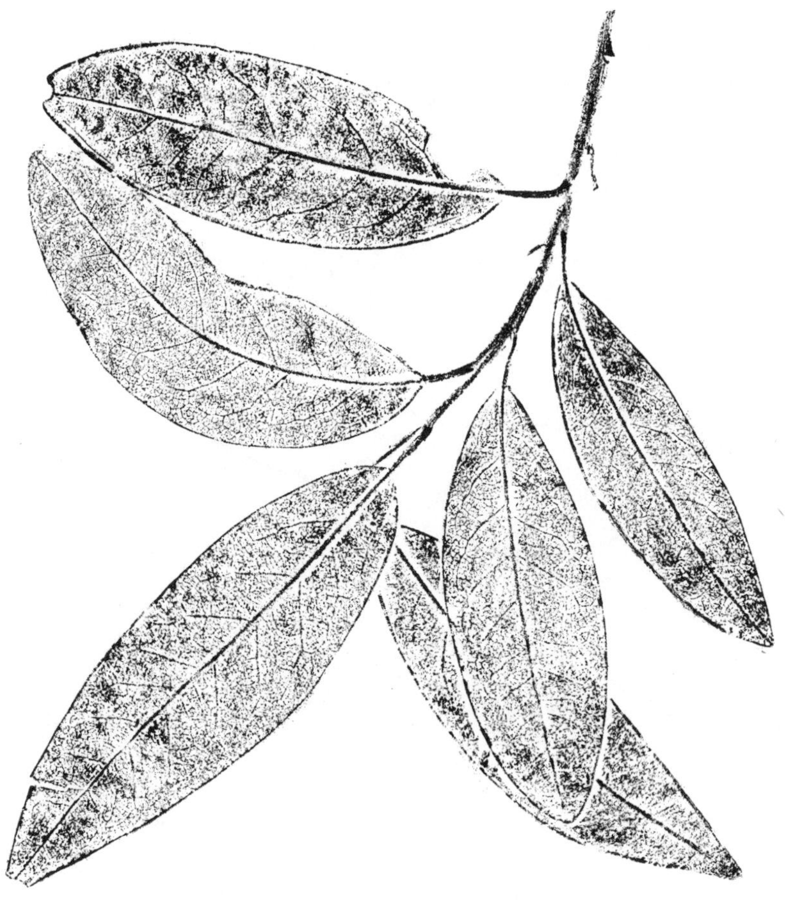

CALIFORNIA BAY-LAUREL
Umbellularia californica

Aromatic leaves, yellow flowers

MAPLE FAMILY *Aceraceæ*
Deciduous

BIG-LEAF MAPLE
Acer macrophyllum

Leaves can be 12–15 inches broad

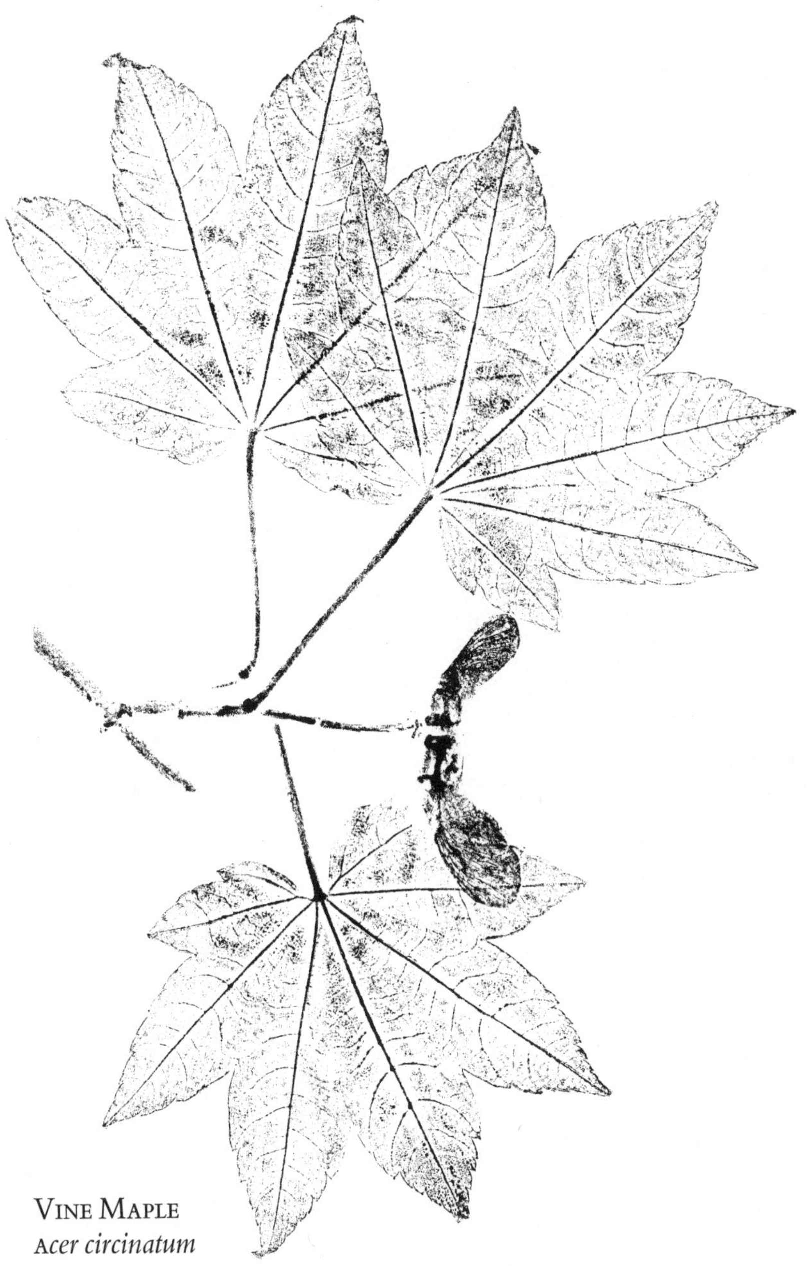

VINE MAPLE
Acer circinatum

Small tree, often vine-like, along northern coast; scarlet samaras with widespread wings

CALIFORNIA BOX ELDER
Acer californicum

Showing flowers and immature leaves; also samara

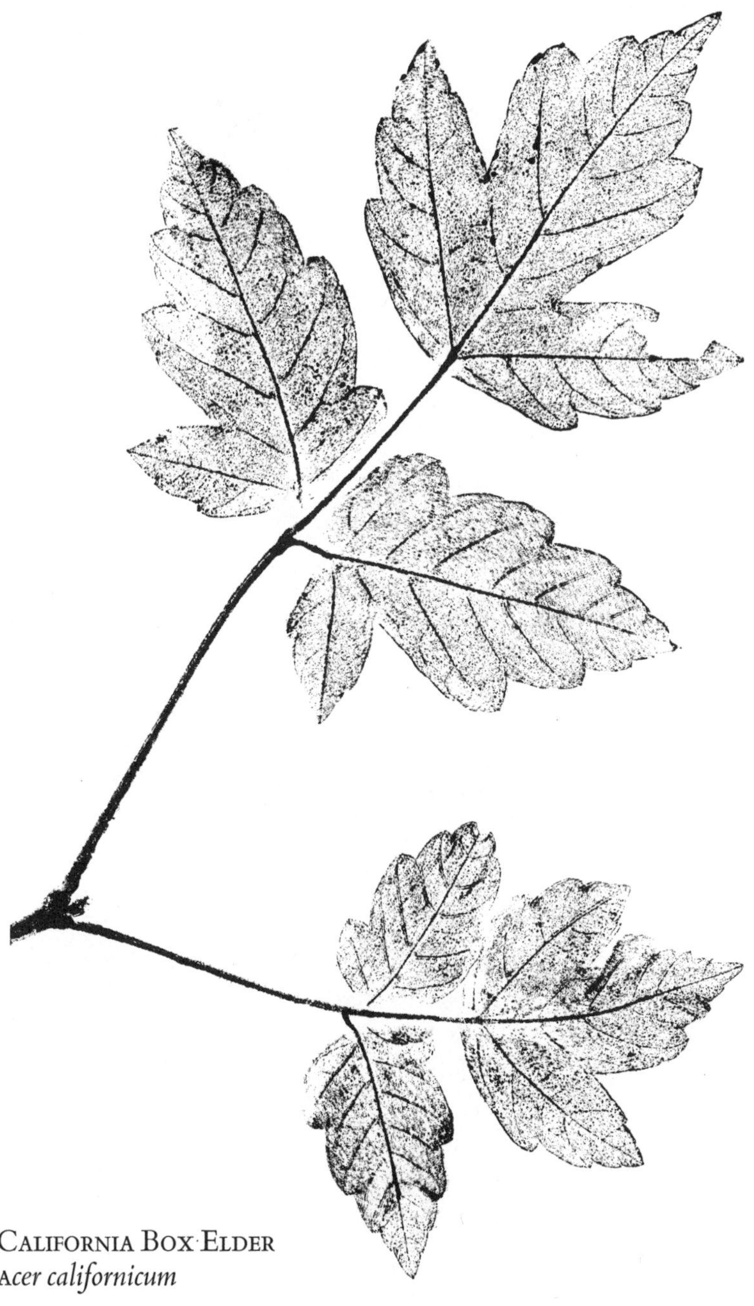

California Box Elder
Acer californicum

Mature leaves of this streamside tree

HORSECHESTNUT FAMILY *Hippocastanaceæ*

CALIFORNIA BUCKEYE
Aesculus californica

Deciduous, leaves palmately compound; shown with flowers

DOGWOOD FAMILY *cornaceæ*

PACIFIC DOGWOOD
cornus nuttallii

White flower has 6 petals

Creek Dogwood
Cornus californica

White flowers, deciduous; along streams

HEATH FAMILY *Ericaceæ*

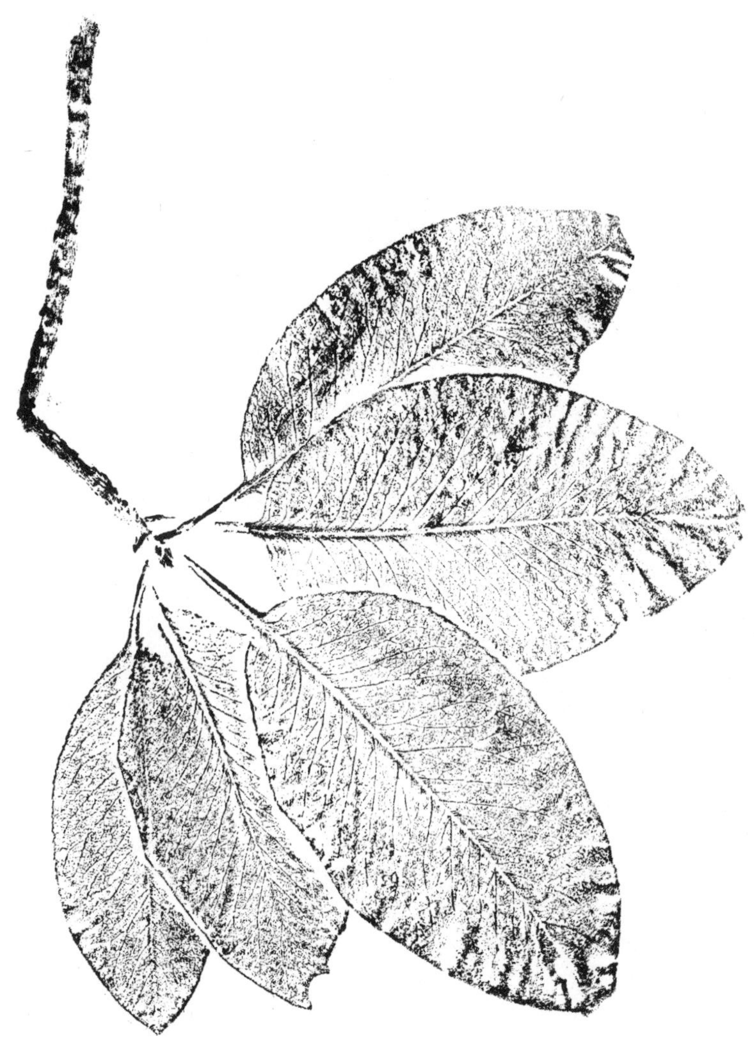

MADRONE
Arbutus menziesii

White flowers, orange berries

OLIVE FAMILY *oleaceæ*

Oregon Ash
Fraxinus latifolia

Deciduous, beside streams

A Note on Plant Prints

THE PLANT PRINTS used to illustrate *The Leaf Book* and my first book, *Marin Trails*, also various magazine articles and a weekly newspaper column, are made directly from the plants themselves. The method consists of first inking the plant and then taking an impression from it with rice paper. The materials needed for making plant prints are block-printing ink, either water soluble or oil based, a medium-sized inexpensive brush, not too soft, and rice paper. These can be purchased in most art stores.

First place the plant on spread-out newspapers on a work table, brush it evenly with ink, keeping the brush fairly dry to prevent puddling, and then move the inked plant to a clean place on the newspaper. Place a piece of rice paper over the inked plant, hold both down firmly, and press and rub the paper gently to transfer the ink from the plant to the rice paper. Lift the paper, turn it face up, and allow the print to dry. Black ink and white paper, in small sizes, are best to start with. Prints can be made in large sizes, however, and in combinations of colored woodblock ink and colored paper.

Most flower petals are thin and fragile, as are also the leaves of many shade-growing plants, and they may not survive more than one print, if that. Sturdy plants like evergreens and grasses often hold up through numerous printings and may be inked and printed again and again. Most plants can be held a few days in closed plastic bags in a refrigerator. Some plants are more amenable to printing after being pressed flat in a thick book over night.

Since I began making plant prints I had always thought the method was in the Japanese fish print tradition, believed to be about one hundred years old. Recently, however, Paula Dawson of Mill Valley sent me an article called "Nature-Printing," in the April, 1971, issue of Garden Journal, published by the New York Botanical Garden and written by Frank J. Anderson, assistant librarian at the Botanical Garden. From it I discover that Leonardo da Vinci printed a leaf of sage with a pigment composed of lampblack or of white lead mixed with oil. The specimen is included in the Codex Atlanticus, created sometime around 1490, and the leaf looks something like our California pitcher sage.

In 1733, the article continues, a Dr. Johann Hieronymous Kniphof in Germany began issuing a "living herbarium" of 1200 examples of plant life. His method was to ink a specimen, insert it into a printing press, and there bring it into contact with a sheet of paper. How many plants he must have mashed in the process I can't bear to think about. Some flowers I work with disintegrate just from the ink alone. But again, before photography, nature-printing provided a more accurate method of recording a plant than some of the botanical drawings of the day, and after seven years Dr. Kniphof's magnum opus was finally completed and published.

Around the same time Benjamin Franklin also made nature prints, using willow leaves and blackberry leaves, and again, like Leonardo, sage leaves, on the notes of Pennsylvania currency. These were used as a safeguard against counterfeiting and no one knows to this day how one leaf could have lasted through a complete edition. Franklin may have made a plaster mold of the original leaf, some think, but it was a secret then and so it has remained.

Also in the 18th century the natives of Tahiti were making tapa cloth by dipping flowers, leaves and ferns in a red sap dye and pressing the plants against cloth to form a design. No one knows how old this method may be – it may predate Leonardo.

Then in the late 19th century photography and the half-tone process were developed and this was followed by color photography; nature-printing as a way of duplicating the natural appearance of a plant fell into a great decline.

Recently, however, the art seems to be reviving somewhat. A group of natural history teachers in Marin County are currently teaching plant printing, both as botany and also as an art form.

I became interested and began to experiment with the method when I couldn't learn photography, even with David Cavagnaro nature photographer, as my teacher. From time to time I have taught plant printing also.

About two years ago, like the Dr. Kniphof I still had not heard about, I embarked on a "living herbarium" I called The Leaf Book.

History seems to repeat itself – but I am happy for an earlier completion than the good doctor's seven years.

A Note on Seaweed Pressings

SEAWEED PRESSINGS are made by floating the seaweed in a shallow pan of water and slipping a piece of drawing or typing paper under your specimen. When it has opened out on the paper and looks well arranged, tip the paper from side to side to drain off the water. The specimen can be opened further with a camel hair brush or eyedropper.

Let the pressing dry a little, then cover with wax paper and put between layers of newspaper. This can be stored in an old telephone book but the newspapers should be changed every day for a week or so until the pressing is dried well and pressed flat.

Seaweed has its own natural mucilage and glues itself while drying to the paper. If not, it can be held down later with white glue.

The finer, lacier marine algae make the best pressings. They are often found washed up on the sand at the drift line, still attached to the coarser seaweed they characteristically grow upon.

When collected in a plastic bag they can be stored in the bag a few days in the refrigerator until used.

NOTES

NOTES

Bibliography

ABRAMS, LEROY. 1940–1960. *Illustrated Flora of the Pacific States.* Stanford: Stanford University Press. 4 volumes.

ANDERSON, EDGAR. 1969. *Plants, Man and Life.* Berkeley: Univeristy of California Press.

BAILEY, L. H. 1933, 1963. *How Plants Get Their Names.* New York: Dover Publications, Inc.

BAKER, H. G. 1965. *Plants and Civilization.* Belmont, Calif.: Wadsworth Publishing Co.

BALLS, EDWARD K. 1962. *Early Uses of California Plants.* Berkeley: University of California Press.

California Native Plant Society *Newsletter*: Room 202, 2490 Channing Way, Berkeley, California 94704.

BLOSSFELDT, CARL. 1967. *Art Forms in Nature.* New York, Universe Books, Inc.

BROCKMAN, C. FRANK. 1968. *Trees of North America.* New York: Golden Press.

COLE, JAMES E. 1959. *Cone-bearing Trees of Yosemite National Park.* Yosemite Nature Notes. Yosemite National Park, California: Yosemite Natural History Association, Inc.

CONRAD, HENRY S. 1944. *How to Know the Mosses.* Dubuque, Ia.: William C. Brown Co.

DAWSON, E. YALE. 1966. *Seashore Plants of Northern California.* California Natural History Guides: 20. Berkeley: University of California Press.

DOWDEN, ANNE OPHELIA T. 1963. *Look at a Flower.* New York: Thomas Y. Crowell Company.

FERRIS, ROXANA S. 1968. *Native Shrubs of the San Francisco Bay Region.* Berkeley: University of California Press.

——. 1970. *Flowers of Point Reyes National Seashore.* Berkeley: University of California Press.

GLEASON, HENRY A. and CRONQUIST, ARTHUR. 1964. *Natural Geography of Plants.* New York: Columbia University Press.

GRILLOS, STEVE J. 1966. *Ferns and Fern Allies of California.* California Natural History Guides: 16. Berkeley: University of California Press.

GUBERLET, MURIEL LEWIN. 1967. *Seaweeds at Ebb Tide.* Seattle: University of Washington Press.

Hardy, M. E. 1935. *The Geography of Plants.* Oxford: Oxford University Press.

Harrington, H. D. 1957. *How to Identify Plants.* Chicago: The Swallow Press, Inc.

Hedgpeth, Joel W. 1962. *Introduction to Seashore Life of the San Francisco Bay Region.* California Natural History Guides: 9. Berkeley: University of California Press.

Howell, John T. 1949, 1970. *Marin Flora, Manual of the Flowering Plants and Ferns of Marin County.* Berkeley: University of California Press.

Jepson, W. L. 1951. *Manual of the Flowering Plants of California.* Berkeley: University of California Press.

——. 1923. *The Trees of California.* Berkeley: Sather Gate Bookshop.

——. 1935. *Trees, Shrubs and Flowers of the Redwood Region.* San Francisco: Save-the-Redwoods League; revised edition, Ida Geary, ed. 1966.

——. 1935. *A High School Flora for California.* Berkeley: Associated Students Store, University of California.

Ketchum, Richard M. 1970. *The Secret Life of the Forest.* New York: American Heritage Press.

Kirk, Donald. 1970. *Wild Edible Plants of the Western United States.* Healdsburg, California: Naturegraph Publishers.

Lawrence, George H. M. 1955. *An Introduction to Plant Taxonomy.* New York: The Macmillan Company.

Leopold, Aldo. 1968. *Sand County Almanac.* New York: Oxford University Press.

Mason, Herbert L. 1957. *A Flora of the Marshes of California.* Berkeley: University of California Press.

McMinn, Howard E. and Maino, Evelyn. 1956. *An illustrated Manual of Pacific Coast Trees.* Berkeley: University of California Press.

Meeuse, B. J. D. 1961. *The Story of Pollination.* New York: Ronald Press.

Metcalf, Woodbridge. 1959. *Native Trees of the San Francisco Bay Region.* California Natural History Guides: 4. Berkeley: University of California Press.

——. 1968. *Introduced Trees of Central California.* California Natural History Guides: 27. Berkeley: Univ. of Calif. Press.

MUNZ, PHILIP A. 1961. *California Spring Wildflowers*. Berkeley: University of California Press.

——. 1962. *California Desert Wildflowers*. Berkeley: University of California Press.

——. 1964. *Shore Wildflowers of California, Oregon and Washington*. Berkeley: University of California Press.

—— and KECK, D. D. 1959. *A California Flora*. Berkeley: University of California Press.

MURPHEY, EDITH VAN ALLEN. 1959. *Indian Uses of Native Plants*. Fort Bragg, California: Mendocino County Historical Society.

ORR, ROBERT T. and ORR, DOROTHY B. 1962. *Mushrooms and Other Common Fungi of the San Francisco Bay Region*. California Natural History Guides: 8. Berkeley: University of California Press.

PARSONS, MARY ELIZABETH. 1955. (First copyright 1897). *The Wild Flowers of California*. San Francisco: California Academy of Sciences.

PEATTIE, DONALD CULROSS. 1953. *A Natural History of Western Trees*. Boston: Houghton Mifflin Company.

——. 1939, 1966. *The Flowering Earth*. New York: Viking Press.

PORTER, C. L. 1959. *Taxonomy of Flowering Plants*. San Francisco: W. H. Freeman & Co.

RAVEN, PETER H. and CURTIS, HELENA. 1970. *Biology of Plants*. New York: Worth Publishers, Inc.

ROGERS, WALTER E. 1965. *Tree Flowers of Forest, Park and Street*. New York: Dover Publications.

ROWNTREE, LESTER. 1948. *Flowering Shrubs of California*. Stanford: Stanford University Press.

——. 1936. *Hardy Californians*. New York: The Macmillan Co.

SHARSMITH, HELEN K. 1965. *Spring Wildflowers of the San Francisco Bay Region*. California Natural History Guides: 11. Berkeley, University of California Press.

SMITH, ARTHUR C. 1960. *Introduction to the Natural History of the San Francisco Bay Region*. California Natural History Guides: 1. Berkeley: University of California Press.

SMITH, GLADYS L. 1963. *Flowers and Ferns of Muir Woods*. California: Muir Woods Natural History Association.

SPENCER, E. R. 1940. *Just Weeds*. New York: Scribner's.

STORER, TRACY I. and USINGER, ROBERT L. 1968. *Sierra Nevada Natural History*. Berkeley: University of California Press.

SUDWORTH, GEORGE B. 1967. *Forest Trees of the Pacific Slope*. New York: Dover Publications, Inc. Originally published by the U.S. Forest Service, Department of Agriculture, 1908.

SWEET, MURIEL. 1962. *Common Edible and Useful Plants of the West*. Healdsburg, California: Naturegraph Publishers.

THOREAU, HENRY DAVID. 1950. *Walden and Other Writings*. New York: Random House, Inc. The Modern Library.

TIERNEY, ROBERT J. et al. 1966. *Exploring Tidal Life along the Pacific Coast*. Berkeley: Tidepool Associates.

U. S. Department of Agriculture. 1949. *Trees, The Yearbook of Agriculture*. Washington, D.C.: U. S. Government Printing Office.

WATTS, T. 1963. *California Tree Finder*. Berkeley: Nature Study Guild.

WENT, FRITS W. and The Editors of LIFE. 1963. *The Plants*. New York: Time Incorporated, Life Nature Library.

WILSON, CARL L. and LOOMIS, WALTER E. 1967. *Botany*. New York: Holt, Rinehart and Winston.

ADDITIONAL BIBLIOGRAPHY

Index